Synthesis Lectures on Chemical Engineering and Biochemical Engineering

This series publishes short books on all aspects of chemical engineering, covering the analysis or design of chemical processes to effectively convert materials into more useful materials or energy. The books will focus on fundamental aspects necessary for chemical engineering design including chemistry, math, physics, and sometimes biology to improve the quality of life by inventing, optimizing, and economizing new technologies and products.

Qianlin Zhuang

From Coal to Hydrogen

A Long Journey of Energy Transition

Springer

Qianlin Zhuang
Kingwood, TX, USA

ISSN 2327-6738   ISSN 2327-6746 (electronic)
Synthesis Lectures on Chemical Engineering and Biochemical Engineering
ISBN 978-3-031-55585-5   ISBN 978-3-031-55586-2 (eBook)
https://doi.org/10.1007/978-3-031-55586-2

This Springer imprint is published by the registered company Springer Nature Switzerland AG
The registered company address is: Gewerbestrasse 11, 6330 Cham, Switzerland

Paper in this product is recyclable.

Preface

Facing the dilemma of the greenhouse gas-derived climate change, there seems no questions about the energy transition that moving into a low-carbon primary energy portfolio globally would be the right thing to do. What remains hotly debated and unsettled, however, is how far the de-carbonization should go, how to achieve it, and how soon to get there. On top of all these, the action over the greenhouse gas-centered climate change calls for a collaborative effort of all who lives on earth, which is why the United Nations Climate Change Conference had just finished its COP28 in Dubai on December 12, 2023. According to BP, the objectives of a zero emission versus a net zero emission certainly have different consequences and may well require different approaches. What solar and wind energy may help Germany achieve either of the objectives would not work in France whose nuclear energy has been its strength and focus in its energy policy. Then, there exists a wide division between de-fossil energy versus de-carbonization, which requires drastically different solutions. For regions and countries that relied on fossil energy or simply on coal achieving either of the objectives would be a long journey with no light in the end of the tunnel.

Practically speaking from historical perspectives, the viable solution to build a de-carbonized or a hydrogen world would not be a simple one, nor would it be perfect. There has to be a host of technologies working hand in hand. The difference between the green hydrogen and the gray hydrogen is only the source of energy to produce them. Scientifically, the evolution of gasification over the past 230 years can be boiled down to the making of the 'water gas' and how to make the most value out of water.

Different from electrolysis to split water molecules with electrons generated from solar or wind into hydrogen the green energy, and oxygen, water gas in a gasification environment is made by "splitting" water molecules thermally by reacting steam with carbon to form hydrogen and carbon monoxide, which the latter could further reduce additional steam to give off more hydrogen, the gray or blue hydrogen depending on if carbon dioxide is contained without being released to the atmosphere. The hydrogen of gray and the hydrogen of green are chemically the same except for utilizing the different characteristics of the redox cycle of water, electric or thermal. The first law of thermodynamics dictates that no matter which way to proceed, it requires the same amount of energy to break up two water molecules into two molecules of hydrogen and one molecule of oxygen. While electrolysis is to assert electricity over water molecules directly the gasification of hydrocarbons such as coals and other fossil materials is to leverage the thermal heat provided by the partial oxidation of the hydrocarbons with oxygen to reduce water molecules into hydrogen while carbon monoxide, carbon dioxide, and some impurities are formed as well. Gasification is an inherently integrated chemical process that has so far provided a large quantity of hydrogen competitively in the current energy market. More importantly, carbon dioxide in the gasification system can be captured as a concentrated stream.

There is a Chinese idiom saying "水火不相容", literally meaning that water and fire can never cohabit, which is often used to describe an inflammable or hostile relationship between two individuals, two parties, or two countries. This is reflected in the ancient philosophies of both the West and the East that water and fire would not be transmutable; water would destroy a fire. As a metaphor philosophically, coal gasification or gasification, in general, is about finding ways to make the *water* and the *fire* not only cohabit but also prop and infuse with each other so as to become or make the valuable hydrogen. In essence, gasification is about the effective utilization of the *fire*, the reactions between *carbon* and *oxygen* to generate the required heat, to mingle with *water*, and to transform *them* into hydrogen and carbon monoxide. Under certain conditions, the more *water* converted the more efficient the gasification would be to reduce carbon footprint in making hydrogen. In this regard, gasification technology does appear to have a lot more work to do from innovation perspective facing the new era of a hydrogen economy. What has been accumulated about science, chemistry, human engineering, and operational experiences over the past 230 years are still fresh and relevant, which should certainly help the gasification technology move to the next level for the production of water gas or blue hydrogen more effectively and efficiently.

Looking back, water has everything to do with the making of coal gas and water gas. During the early days of coal gas making, water generated from the decomposition of the volatile matter contained in coals left a retort without being utilized and ended up in a water liquor, the ammoniacal liquor, which has to be treated with additional efforts and capital. Then in order to make use of the red hot coke discharged from a retort operation, interested engineers and inventors attempted to recoup the heat from the residual red hot coke by spraying water on it for more coal gas but only found out that the benefits were hardly justified by the efforts. Such trials, however, led to the change in chemistry for coal gas making. When the Siemens brothers invented the open hearth process in 1865 they placed a water trough right beneath the grate of the gas producer primarily to protect the grate by drawing heat from the hot ash clinkers. The naturally evaporated steam rising up from the trough brought up some unexpected benefits by reacting with carbon both in the producer and in the regenerator, giving off additional inflammable gases, which is very valuable to further help achieve the desired temperature in subsequent combustion for steelmaking. In 1879, in order to open up the internal combustion engine market, Dowson set up a small coal-fired boiler to turn water into superheated steam and used the steam to make a quality producer gas to fire up gas engines of as small as a few horsepower. The superheated steam would also assert a positive pressure to the gas producer, which is necessary to maintain a normal operation of the compact system. Dowson did it well by riding right into the internal combustion engine market, which was further expanded by Mond to feed producer gas to much larger gas engines, which had continued well into the twentieth century. Although getting into the gasification business might have been an unexpected consequence for Mond whose original intention was simply to fix Air-N to make ammonia. Mond created a gasifier operated by utilizing a large amount of water, of course in the form of steam, to "extract" the tiny amount of coal-N into ammonia. Such a high steam condition created a gasification environment allowed him to maximize the recovery of ammonia to manufacture the most wanted fertilizer. Such a small amount of fertilizer, nevertheless, had made his business a success by contributing more than 10% of the overall sales revenue. Producer gas typically contains about 50% of nitrogen because of the use of air blast, resulting in a low BTU gas, which is better suited as a fuel gas. Now back in the United States in the 1870s, the Civil War veteran Lowe made an ingenious invention to allow him to manufacture the first water gas or syngas containing only a small amount of nitrogen. To be able to do so, Lowe resolved the long time struggle with gasification technology in terms of effectively providing the required heat to support the reactions between carbon and steam, which is highly endothermic. With the required heat generated from the air blast cycle, Lowe was able in the steam run cycle to generate a high-quality water gas, primarily hydrogen and carbon monoxide. This water gas then in the next century made the Haber-Bosch process an industrial miracle. Equally important in producing hydrogen is the CO catalytic converter or the catalytic shift process that BASF invented, which made the water gas an indispensable raw material for the production of liquid fuels in Germany before the end of WWII.

The list goes on and on but would dry out without water involved. Each of the cases also carries many solid stories about the unique combination of passion, curiosity, ingenuity, perseverance, resourcefulness, and an industrial mindset that are all important to bring an idea into an industrial reality. Then fast forward to the 1950s, a time when innovations appear to become institutionalized, Texaco Inc. leveraged its DNA as an oil company to use, again, the water to turn solid coal fines into a flowing liquid that can be conveniently pumped into a pressurized entrained gasifier, which pretty much removed the one of last hurdles before establishing the modern gasification technology. The modern gasifiers deployed today can be designed with a unit capacity of as high as 3,000 tons of coal daily and even heading toward the level of 4,000 tons. The resulting scale economy along with additional advancement in materials, engineering, and operational expertise have greatly helped enhance the competitiveness of today's entrained coal gasification technology.

Moving forward, the increasing human economic activity would most likely continue its uptrend, which would demand more energy to foster the increasing human activity. To counter the potential uncertainties around the subsequent impact on climate change and the biodiversity of the global ecosystem, gasification technology as one of the best available and effective technologies having a solid track record should be well positioned to help manage the carbon footprint of the future energy market considering the current reliance on fossil energy. It would also be part of an insurance policy for the needed transition to the future de-carbonized world. Considering the long journey that the gasification technology has endured and evolved the future of the energy transition would less likely be a short one, and an insurance policy would certainly help minimize any unforeseen risks. During its evolution over the long past, that the gasification technology has been able to transform and recreate itself at almost every turn along the industrial revolution and to become more resilient and efficient every time should provide credible hopes for gasification to make another transmutation from "水火不相容" to "水火相容". Facing the reality as a semi-empirical science, in the meanwhile, the innovation of coal gasification technology would certainly have to continue in order to extract the most value out of water. Decoding the ultimate chemistry and related dynamics on a molecular level that governs the reactions inside a gasifier would be necessary as well. Fundamentally, what hydrogen and oxygen did 250 years ago to help Lavoisier decipher the secret code behind the phlogiston theory and establish the first chemistry, on which gasification is based, seems still inspirational. This time, however, three elements instead of two, hydrogen, oxygen, and carbon would have to work together, which may guide us, at least philosophically, to make another leap forward in advancing the needed science and technology. Such a fundamental knowledge would in turn lead to an industrial solution that would allow the three elements not only to cohabit but also prosper and make the blue hydrogen better and more efficient, which is the essence of the "水火相容", which would contribute to establishing a sustainable green future of the energy transition along with other renewable technologies. In addition, looking at the potential synergies between renewable energies (wind and solar) and coal gasification there seem some potentials that would warrant additional exploration to further shrink the carbon footprint of making blue hydrogen.

Well, how likely would the world of net zero or zero carbon emissions be achieved? The answer may be that we don't know until we try it out. For gasification, it has been a long journey, a journey full of hard work, challenges, opportunities, innovations, excitements, and rewards. What lies ahead would unlikely be a short and smooth one either but would certainly be full of surprises and hopes.

Kingwood, TX, USA   Qianlin Zhuang
                                                                                            zenrinshou9@gmail.com

Contents

Introduction

Coal gasification is a process to convert coal into coal gas, instead of direct combustion. From a historic perspective, it is a collective body of coal gas making processes that convert coals into coal gas of different types that well served a few times of energy transition over the past 230 years, especially during the Industrial Revolution when there were no other alternative fuels available. Coal gas is primarily made of hydrogen, carbon monoxide and carbon dioxide, and a small quantity of others including methane, nitrogen, hydrogen sulfide, ammonia, olefins, and other hydrocarbon gases etc. The composition of a coal gas changes depending on a specific coal gas making process while further impacted by the conditions under which the coal gasification takes place. Then, the coal gas making process changes, evolves, adapts, and reinvents itself when necessary at times in order to survive or to better serve the different needs and purposes during the course of the industrial revolution and thereafter. Along the way, the coal gas making has taken different forms such as a horizontal retorting, gas producer, regenerative retorting, water gas generator, vertical retorting, slagging gasifier, fluidized bed gasifier, fixed bed gasifier, and entrained bed gasifier etc. Coal gas making or coal gasification process has been advancing by replacing the old with a new in the long run, but more often at times many of them would cohabit and work hand in hand to meet the unique needs of the time, which have made the industrial revolution much profound and consequential in terms of not only the advancement of science and technology but also the progress of social economic development. To give a few examples, coal gas first illuminated the increasingly crowded cities and towns when the world was in a desperate search for an alternative source of lighting to candles and oil lamps. Coal gas also enabled the creation and commercialization of internal combustion engines or the gas engine that had led to a series of significant industrial developments, which is still an important part of

Q. Zhuang, *From Coal to Hydrogen*, Synthesis Lectures on Chemical Engineering and Biochemical Engineering, https://doi.org/10.1007/978-3-031-55586-2_1

modern lives. Another coal gas, the producer gas, unleashed the constraints limiting the mass production of steel when the world was in urgent need of steel for the construction of train tracks and high skyscrapers. Then, it is the water gas that not only extended the era of coal gas lighting once being carburetted with a light oil but also made the hydrogen available for the first fertilizer that was artificially manufactured, one of the most important commodities in human civilization that have ever since supported the fast growing population worldwide. The list goes on and on. If it is generally accepted that there would be no industrial revolution without steam engines, it would not be unfair to say that the industrial revolution could have been much shorter and far less influencing should there be no coal gas or coal gasification. Metaphorically, if coal runs the blood circulation system of a human body by providing it with the most needed energy to keep human organs, tissues, and cells running, then, coal gas would provide, in addition to the energy, the system with the most needed and necessary nutrients that human organs, tissues and cells require to function properly and healthily.

From the onset of the industrial revolution, coal had indeed become the only source of viable energy without which the industrial revolution in Britain would not be able to take its root. With the industrial revolution rolling on, such a thirst for energy was so dire that the Newcomen steam engines with an efficiency of only 0.5% were plugged in to drive water pumps to expedite the coal mining operation around Britain during the early 1700s. Soon, coal had become the bloodline to keep the industrial machines up and running during the nineteenth century and the centuries that followed. Such a status quo has ever since remained essentially unchanged, but only to a lesser degree by percentage. In 2020, our world consumed 3.7 billion tons of oil equivalent coal, which was 28% of its primary energy use overall. Along with crude oil and natural gas, fossil energy collectively contributed 85% of primary energy worldwide for the same year (***data from BP website***). Such a large consumption of fossil energy, the heightening public environmental awareness in the past several decades, and the recent debates over climate change of both scientific and political have, therefore, inevitably brought the fossil energy to the fore. From the perspective of climate change, it is undisputed that coal releases the highest carbon dioxide and other pollutants like SO_x, NO_x, mercury, and ash etc., when burned, due to its high carbon content and impurities inherited from its long time geological formation. Coal fired power plants have become the single largest source of CO_2 emission noticing that generating one MWh of electricity would release about 750–850 kg of CO_2 to atmosphere, which depends on the technology of the generation fleet and the coals used. For the production of fertilizers and chemicals from coal, one ton of ammonia or methanol would emit up to 3.8 tons of CO_2, and one ton of polyolefin from bituminous coal would result in about 10 tons of CO_2. All these sources have contributed to the CO_2 buildup in the atmosphere. Historically, CO_2 would seem to be the obvious culprit by looking at the increasing human activities amidst the frequent happening of natural disasters such as extreme weather, hot and cold, hurricanes, and flooding that have caused tremendous tolls on our economy and human lives. Such an increased public awareness over the potential impact of the emission of carbon dioxide has logically pointed to the continued and increasing consumption of fossil energy, and coal would be the first to blame.

Speaking of coal, it is a long story of love and hate. The fact that coal is an inferior fuel, the fuel for the poor, has been known for a long since the time of the Roman Empire. There have been times that coal burning was banned or restricted due to the black smoke and its pungent smell emitted from coal burning. The problem, however, is that its abundance, convenience to access and handle, and its low cost make coal an irresistible source of energy to meet the ever increasing industrial activity every day. This is especially true when the Industrial Revolution took off in Britain, then in Europe, and North America that coal has become the irreplaceable source of fuels in order to power the thousands of industrial steam engines, locomotives, steam boats, and oceangoing ships and so on. During the most of the nineteenth century when the same smoke, black, greasy and pungent, re-emerged from coal burning in cities like London, Birmingham, Manchester, New York, Philadelphia, Frankfurt, Hanover and so forth, banning the use of coals then simply became a hopeless action. Instead, local authorities, publics, and companies had turned to coal gas to eliminate those chimneys spewing the black smoke. Steam engines were replaced with coal gas fired gas engines and coal fired metallurgical furnaces, industrial boilers, heating, and cooking etc. were switched to coal gas as fuel. It worked and coal gas as a clean fuel had achieved a wide recognition. Such a status quo of the coal gas as the preferred fuel continued through the two wars and finally gave way to natural gas or petroleum gas after the WWII in the US and about the 1960s in Britain and Europe. Coal, nevertheless, has remained its significance as a primary energy as of today. This is especially true for countries that lack reserves of crude oil and natural gas but are rich in coal reserves or have access to it. The simple truth is that coal in those countries will continue to play a pivotal role to foster the social and economic development. With emissions of SOx, NOx, VOCs and particulate matters well managed from the combustion of coal by applying the best available technologies, the concerns with regard to the use of fossil energy especially coal appears to hinge on the large amount of carbon that coal is inherently made of. De-carbonization has been pursued or targeted either as a vision toward developing a future carbon-free hydrogen economy or by harvesting more renewable energies from Mother Nature such as wind, solar, biomass, and wastes. Interestingly, it is coal 260 years ago that set the industrial revolution free from Mother Nature first with steam engines and then with coal gas engines. Now, Mother Nature has circled back, enabled by a series of new technologies developed in recent decades such as solar panel making, battery storage, and wind turbines and so forth. Would it be possible for Mother Nature to free this world from using coal or fossil energy this time around? Or, would Mother Nature have to work with coal and coal gas in order to build a de-carbonized energy future? There seems hardly to be a definite answer, but the answer would certainly be dependent on what is available on the table in a pragmatic and foreseeable way. From where we have come from so far, however, it appears that fossil energy including coals would stay in the picture for sure and for a while, and that innovation and technology would hold the golden key to the de-carbonization of the fossil fuels and the recycle of the captured carbon.

Today, hydrogen seems to become one of the highly sought energy sources that would help fight against the climate change and facilitate the ongoing energy transition to another level. In fact, hydrogen, ever since discovered by Henry Cavendish in 1766, had been a gas of essentially no value and, therefore, did not receive much attention until the arrival of the synthetic age 147 years later. Except the fact that hydrogen along with oxygen discovered by Joseph Priestley in 1874 provided the French chemist Antoine Lavoisier the necessary evidence to establish modern chemistry, hydrogen went unnoticed into the millions of burners of coal gas and producer gas. Although being a major component that makes coal gas hydrogen has no illumination value when coal gas was first deployed to light up streets and houses; it was rather a component of concern in the producer gas when being used in large gas engines at the end of the nineteenth century due to its tendency of premature ignition that could cause an engine misfire. As a fuel gas, hydrogen indeed is a green energy source because its combustion gives off only water vapor. Subject to how it is made, however, it may well be a different story.

In principle, hydrogen can be produced electrically and thermally, which both ways are represented by technologies of more than two hundred years old. There is the electrolysis still under development to split water with electricity, which currently seems promising due to the advances made in the renewable energy of solar and wind. Current industrial scale hydrogen, however, have been manufactured thermally via technologies such as gasification, steam methane reformer (SMR) and autothermal reformer (ATR) etc., among which gasification is the oldest state of the art way before SMR and ATR were born. Gasification as an extension to coal combustion turns coal into coal gas that takes different forms has played critical roles along the industrial revolution. In addition, prior to the first artificial ammonia, coal gas was also the largest source of ammonia for the manufacture of ammonia sulfate as fertilizer and as an important raw material for the Solvay process to make soda as well during the late 1800s. Targeted as one of the best available technologies for the generation of electricity, in the meanwhile, gas producer was tied with gas engines, large and small, to generate electricity, which becomes the early version of the integrated coal gasification combined cycle (IGCC), which has been demonstrated in the past 30 years worldwide.

From the viewpoint of hydrogen, the advent of the water gas generator is a significant milestone in the course of coal gasification; the invention rendered the resultant coal gas with both high hydrogen content and a high quality, which is also called syngas. Modern coal gasification technology has continued to evolve along the trend, manufacturing syngas in a large scale that is conveniently conditioned to hydrogen, carbon monoxide, or a mixture of both as final products. When BASF successfully used hydrogen to fix air nitrogen over an alumina activated iron catalyst to manufacture ammonia in 1913 at Appau, hydrogen suddenly becomes one of the important industrial commodities as feedstock for chemical purposes. When integrated with a several downstream process technologies developed, coal gasification has become one of the most effective processes to de-carbonize coal or other fossil materials to produce hydrogen, one of the largest sources of high purity hydrogen.

After providing hydrogen to the first synthetic ammonia, coal gasification continues to evolve by manufacturing the needed hydrogen and syngas for other synthetic processes such as the first methanol synthesis in the mid-1920s, the first aviation and motor gasoline synthesis in the mid-1930s, and then, in the past thirty years, a wide range of valued added chemicals such as high alcohols, acetic acid, glycols and olefins etc. Today, coal to chemicals has become a viable alternative of chemical supply complimentary to the crude based petrochemical industry in countries or regions depending on coal to feed their economic growth. In China, coal gasification converted at least 170–200 million tons of coal each year to fertilizers, chemicals, liquid fuels and other products based on the built capacity of coal gasification prior to the COVID breakout. Coal gas making or coal gasification in its primitive form during the early days provided by-product coal tar as raw material for dyestuff making.

In retrospect, coal gasification has taken a long journey in its commercial development and deployment including related processes and the corresponding knowledge buildup to where it is today. The coal gasification process itself has transformed from a retort, gas producer, blue water gas generator, carbureted water gas generator, fluidized bed process, fixed/moving bed process to today's entrained bed gasification process; its operation evolves from a small scale and batch operation to a large scale and continuous one; its operating conditions change from low temperature, atmospheric operation to high temperature and high pressure one; the use of product gases, driven by the markets at different times, shifts from illuminating purpose, heating for industrial processes and buildings, small engine primer fuel, automobile fuels, chemical feedstock, and to electricity generation; its feedstock portfolio has expanded from coal to coke, from good quality coals to poor quality ones, and back and forth impacted by either market conditions or coal access at different times. The roadmap of coal gasification has been up and down from time to time because of changes in the energy market and economic conditions.

This writing provides a detailed overall lookback at the evolution of the major coal gas making processes, their industrial deployment, and the resulting impacts to the industrial revolution. Author also made attempts to demonstrate the interaction between coal gasification technology and chemistry through industrial examples and to further explore how the interaction inspired coal gasification technology to advance to where it is today. This writing, however, is not intended to be a comprehensive review of coal gasification as a whole as it would be too broad to cover. In many ways, the evolutionary progress and the achievements of coal gasification reflect the contradictions and challenges around the use of fossil energy to social and economic development, which seems to be an ideal epitome of the past and still of practical significance today. Looking back, the use of coal, while as an integral driver of industrialization, has also been the source or cause of countless challenges throughout history. The utilization of coal gas has made the process of industrialization much deep and extensive that has penetrated into almost every corner of public life. From the perspective of the scientific and industrial revolution, this writing takes readers 250 years back when chemistry did not exist to review the inventions, developments and industrial applications of major coal gas making technologies, related key individuals and cases chronologically through the establishment of modern chemistry, the emergence of gas lighting, the breakthrough in steelmaking technology, the invention of an internal combustion engine and its industrial application, the emergence of artificial fertilizers, and major technological breakthroughs in the era of synthetic chemistry. Through these examples, readers should be able to explore how coal gasification had such a profound and far-reaching impact on the industrial revolution. By using many real cases and quotes of those great minds and practitioners, the writing attempts to create the moments of the past so that we could relive and get more sense out of the past in terms of the evolution of coal gasification technology and its working under the context of the industrial revolution and beyond.

As far as coal gasification technology is concerned, its evolution process is full of challenges, no matter whether these challenges stem from its own success or threats from other technologies. Interestingly enough, every challenge that coal gasification technology faced has been transformed into new opportunities through the hard work, scientific discoveries, and technological innovations made by those great minds, and backing out of each transformation time after time again, coal gasification technology would have emerged stronger and more robust. A lookback of the evolution of coal gasification seems necessary and beneficial in the midst of the energy transition centered on the greenhouse gas derived climate debate and the ongoing search for a viable solution to the energy future. After all, coal gasification technology has proven to be one of the effective tools not only to de-carbonize fossil fuels but also to manage the subsequent carbon dioxide well.

Different from any previous energy transitions, current transition centered on climate change and biodiversity faces challenges of a global scale and a live cycle in characters, instead of regional and segmental, which calls for a collective action globally that needs to be addressed with an approach from cradle to grave. It is author's hope that the evolution of coal gasification technology filled with rich experience, excellence, and knowledge with innovation would help us land on a solution long lasting. At least, it will entertain readers with a lot of critical thinking, inspirations, and aspirations as well to today's technological innovation and the search for a road that leads to a sustainable future.

Early Energy Industry

The seventeenth century Europe especially England had experienced a burgeoning commercial development in maritime trading, financing, textile making, etc. As a result, public lives and living standards started to see changes for the better with increasing demands for daily products related to hygiene, clothing, fashion and so forth. At its core, shipbuilding and metallurgy for iron making are two pillars that supported such development. More iron production in turn means more demand for fuels, charcoal and coal, to make the needed iron. That charcoal was a better fuel than coal had been known for thousands of years. During Roman times back in 500 AD London, a charcoal was said to be the fuel for the wealthy who could afford of it and a coal for the poor as it was cheap and polluting when being burnt. The coal burning pollutes every corner of a public quarter, the smoke burdened typically with a pungent and irritating smell due to sulfide laden chemicals and ammonia, the oily waste water due to the formation of tars and oils, and the dusty coal ash piling up the landscape, especially when the burning takes place without any pollution control. History repeats itself from time to time that population growth increases the consumptions of both charcoal and coal; then the balance tends to shift toward more coal use when an over-logging eventually causes deforestation. When pollution becomes out of control some extreme measures would have to be taken. There was such a time in 1306 in England, for example, that King Edward had to install a ban of a coal burning by a death penalty in order to combat the notorious smog issued from the coal burning. Soon, with the industrial revolution on the verge 300 years later the hope to ban coal burning had become hopeless.

Q. Zhuang, *From Coal to Hydrogen*, Synthesis Lectures on Chemical Engineering and Biochemical Engineering, https://doi.org/10.1007/978-3-031-55586-2_2

Coal and Steam Engine

Entering into the seventeenth century England's charcoal shortage for iron making had become severe enough so that it had to rely on the import from other countries such as Sweden, America, Russia and others to meet its domestic demand, which was complicated at the time by the frequent regional geopolitical tension. To manage the potential geopolitical risks, England had to turn to its large indigenous coal resources by opening up more coal mines to support its iron making and shipbuilding industries, which was critical for its maritime commerce at that time. By mid-seventeenth century coal mining had become a widespread operation and most of its coal fields had been opened up, delivering about 2 million tons of coal annually. This number went up to about 3 million tons by 1700 (Smil, 2017). Coal mining had since become a critical industry, which by 1800 supplied 90% of the energy that the country's economy required and employed a large number of workers, old and young, male and female. Still, coal production could not match the growing demand. With the mining operation dug deeper and deeper a frequent water flooding required significant efforts to move water out of a mine shaft. Based on the 1698 invention made by the military engineer and inventor Thomas Savvy (1650–1715), Thomas Newcomen (1664–1729) invented the first industrial steam engine to pump water out of coal mines in 1712. In its primitive prototype the Newcomen engine stood a three story tall and offered an efficiency of only about half a percent. This giant machine, however, did improve the operation of a coal mining and started to add an impact on the industry that gradually led to the industrial revolution. By 1775, there were about 600 such engines built in England.

In Newcomen's early design he used a brass material to build his steam cylinder, which was weak and required a frequent maintenance. In 1720, he was able to upgrade the material to a cast iron, a stronger material made by Abraham Darby (1678–1717), so to make the steam cylinder larger and more endurable. Darby was able to make the cast iron with a better quality at that time because he introduced coke in the iron making at his new establishment in Coalbrookdale, England back in 1709. Before this time, charcoal was typically used for iron making. Darby's introduction of coke into iron making was a significant milestone because the higher energy density and strength of a coke allows a blast furnace to go lager while achieving a higher temperature, which is necessary to produce a more amenable iron. During the early 1700s, Darby's blast furnace only produced about 300 tons of a cast iron a year; 50 years later, a modern day blast furnace could easily produce 15,000 tons of iron daily. 83 years later, the smoke released from Darby's coke making oven or alike would be utilized for a different purpose, that is to make coal gas, an inflammable gas to light up streets and houses, which is the subject of this writing.

The Search for an Alternative Lighting

Ever since adopting the Darby's cast iron to build steam cylinder, the Newcomen's steam engine had remained essentially unchanged for more than 60 years until James Watt (1736–1819) made his famous technical redesign of the Newcomen's steam engine by setting up a standalone steam condenser in 1769. With additional improvements, Watt in 1775 was able to improve the thermal efficiency of the Newcomen's steam engine to 2.5%, five times of the Newcomen's. Such a drastic improvement certainly enhanced the operation of a coal mining by going deeper and bringing more coal out of the coal mines at a lower cost to the market. First time in human history, the steam engine had transformed the power source that was typically obtained from natural, animal and human power to a mechanic power, kicking off the industrial revolution that encompasses a whole series of domino events that led to our modern day. With more cheap coals becoming available the coal gas for lighting would be the next important event to come. (Accum, 1815).

Coinciding with the deployment of the steam engines in England, its textile industry took its root, aided by the authority's ban of both the import of the calico, which was mainly imported from its colonial possessions such as India and others. The increasing demands for fashion and clothing in England accelerated its local textile industry. A series of inventions at home, one after another working in lockstep, had been developed and deployed in the rapidly changing textile industry. A few examples are the flying shuttle invented by John Kay in 1733, the water frame by Richard Arkwright in 1769, the Spinning Jenny by James Hargreaves in 1770, and the Spinning Mule by Samuel Crompton in 1779 and so on. These inventions, aided by the available steam power, had transformed the England's textile industry from a home based individual operation to an institutionalized factory setup. When introduced his power loom in 1786 Edmund Cartwright gave the textile industry the final spur to take off and become a highly mechanized system. In doing so, he brought in steam engines to replace the water power to drive all the wheels around a factory, which also made machines to perform many tasks previously carried out by skilled human hands. With more coal derived steam power becoming easily available, the operation of a textile mill was no longer necessary to be sited next to a river or a creek in order to harvest the water power; many mill owners could now pick up a site convenient either to the resources of materials and labors or to a market whichever made more business sense.

Textile making is a labor intensive field that requires many skillful hands and hard-working humans. With more factories being set up, more people were flocking to the factories to join the labor force; then towns emerged and small towns grew into larger ones. New mill towns like Leeds and Manchester came to existence and grew in a rapid pace; their populations in 1750 were 16,000 and 18,000, respectively, and grew into 53,000 and 90,000 fifty-one years later. Similarly in Birmingham and Sheffield specialized with metal works, their populations grew from 12,000 and 24,000 in 1750 to 31,000 and 74,000, respectively, in 1801. By 1851, England had about 38 cities with a population over 40,000, comparing to 9 in 1801 and only 2 in 1750 (List of towns and cities in England by historical population). By the end of the 1700s England as the world dominating imperial power had already created a sizeable middle class in addition to those wealthy, old and new, who were looking for ways to improve their living standards and working environments. Fashion and clothing had become a popular representative of a power or a privilege; and bathing started to take its regular frequency. The bathing habit as a way to improve hygiene then made soaps into one of the luxury products, which then spurred an industrial development for soda making, the earliest chemical activity at that time. On another front, searching for an alternative lighting to the dim and expensive candle or tallow light had become an urgent event as well. Interesting enough, the scientific discoveries of gases or "air" during the latter half of the eighteenth century that led to the establishment of the modern chemistry certainly paves the road for exploiting the coal gas in lighting up the streets, houses and workshops in the next century, which also kicked off the journey of coal gasification.

To make it short, what the textile industry had achieved especially in the latter half of the eighteenth century was largely benefited from the technology innovations and industrial deployment. Constantly searching for something new to improve its operation, bottom-line or status-quo appears to have embedded into the DNA of the textile industry, at least, in England in order to survive and stand out amid the fierce competition in the market place. When coal gas making technology became available many owners in the textile industry were eager to embrace it to explore the potential to light up their increasingly large and crowded factories.

References

Accum, F. (1815). *A practical treatise for gas-light.*
Smil, V. (2017). *Energy and civilization: A history.* The MIT Press.

Coal Gas and Chemistry

3

Generally speaking, coal gasification is about making gases from coal, which the gases are valuable and can be used for a broad range of applications such as for illumination, heating, cooking, iron and steel making, power generation, and chemical synthesis. Then, knowing those gases and their characteristics would naturally be necessary not only for the development of coal gasification technology and its related process but also for the operation and maintenance where the gases would be applied to. Well, such a way of thinking or logic appears normal when the discipline of applied chemistry has become part of our methodology or DNA. Unfortunately, this is not the way it was way back then. As late as the eighteenth century, chemistry was by large the operating field of artisans, alchemists, and philosophers. The development of a coal gas making technology would be no surprise at all a process of the rule of thumb and a trial by error, a time consuming process. When James Watt filed his 1769 patent of *"A New Invented Method of Lessening the Consumption of Steam and Fuel in Fire Engines"* as an improvement to the Newcomen's steam engine used for water pumping, there was little knowledge existent on what air or gas was. Air was still one of the four ultimate elements under the Aristotelian philosophical view. Actually, like many in England at the time, Watt was a phlogistonist, which is less known than that he is a great inventor of the steam engine. Watt was engaged in many discussions about pneumatic chemistry. Here is what he put in the patent explaining how he improved the Newcomen's engine.

> My method of lessening the consumption of steam, and consequently fuel, in fire engines consists of the following principles: First, that vessel in which the powers of steam are to be employed to work the engine, which is called the cylinder in common fire engines, and which I call the steam vessel, must during the whole time the engine is at work kept as hot as the

Q. Zhuang, *From Coal to Hydrogen*, Synthesis Lectures on Chemical Engineering and Biochemical Engineering, https://doi.org/10.1007/978-3-031-55586-2_3

steam that enters it…secondly, in engines that are to be worked wholly or partially by condensation of steam, the steam is to be condensed in vessels distinct from the steam vessels or cylinders, although occasionally communicating with them. These vessels I call condensers, whilst the engines are working, these condensers ought at least to be kept as cold as the air in the neighborhood of the engines by application of water or other bodies. Thirdly, whatever air or other elastic vapor is not condensed by the cold of the condenser, and may impede the working of the engine, is to be drawn out of the steam vessels or condensers, by means of pumps wrought by the engines themselves or otherwise.

Somehow, Watt might have believed that water was a compound body of *pure air* and phlogiston. While we have no way to know what's in Watt's mind about the *pure air*. The *pure air* or the *elastic vapor* coming from water but not condensing in the condenser could be what would cause his engine less effective. Watt would be right if he meant the pure air the dissolved air in water, which we all know well today that air has a certain degree of solubility in the water depending on conditions. Upon heating the dissolved air would escape. This is why the boiler water in modern engineering practice has to be treated through a DEMIN process to remove the dissolved air and minerals before entering into a boiler water system. Otherwise, the air would build up in the water circulation system to make steam less effective to do work, and the minerals would eventually deposit to form scale in the system, slowing down heat transfer and increasing the burden of maintenance. Considering the primitive nature in design at the time, we can imagine that any leakage from inadequate seals could also contribute to the air in the cylinder and the condenser impacting its effectiveness. Unknowing the air and vapor chemically Watt was clear that the *elastic vapor* and the *pure air* noncondensing but contracting and expanding with pressure on and off, would create a dead space in the cylinder and the condenser, which did not help the engine carry out an effective work. To Watt, this *air* seemed unfathomable and deserved a close look if possible. Understanding that the Newtonian principle on gravity was prevalent in explaining the elastic nature of air and other vapors, the school of thought on chemistry was quite mechanical as well at the time. With progress made in pneumatic chemistry in the next decade, however, the myths of *air*, *airs,* and their chemical tendencies were eventually debunked, which led to the overturns of both the mechanical chemistry and the phlogiston theory prevailing over the previous 100 years.

Coal Gas—The Inflammable Air

Like fire, human beings had been aware of air around us and its critical nature to our lives for a long, and had learned how to use the fire and the air to make tools, wares, and weapons. Until the later part of the eighteenth century, however, the fire remained a mystery and a sacred phenomenon in the hands of a few like religious clerks, artisans and alchemists, and the air went by unnoticed or was simply taken for granted, probably because of its intangible, odorless and colorless nature. This is exactly what Jan Baptista van Helmont (1580–1644), a Flemish (today's Belgium) iatrochemist, demonstrated in his famous but controversial experiment of tree growing around 1640. In his garden, Helmont planted a 5 pound willow tree into a big pot holding 200 pounds of dry soil. By watering only routinely over the next five years, the willow tree grew into a 169 pound one. Based on the information obtained by measuring the weight of the soil in the pot, Helmont concluded that water was all that a tree would need to grow as the soil in the pot remained almost intact. According to Aristotle's philosophical view, air and fire were two of the four basic and ultimate elements (*fire*, *water*, *air* and *earth*) that make up our world, and the four elements transmute from one to the other. To explain the chemical phenomena taking place around daily life such as fire and the decomposition of some salts, Georg Ernst Stahl (1656–1734), a German chemist, introduced a theory of phlogiston to expand the Aristotle's philosophy. According to the phlogiston theory, substances such as coal, wood or metal were made of calx (oxide) or ash and phlogiston. Upon burning or heating, the phlogiston ("*air*") escapes from its body, giving off flame, light and/or heat and leaves calx or ash behind; phlogiston or air would then be fixed back into plants or vegetables, which was what Stephen Hales (1677–1761) tried to demonstrate in his famous treatise, "*Vegetable Statick*", published in 1726. In the work, he found that air somehow was breathed in by vegetables at different parts like roots, trunks, branches, and leaves as food that vegetables would rely on to grow. This probably prompted him to study the nature of the air reversely by cooking the *air* out of the vegetables. Hales spent 210 pages out of his 358 page treatise presenting his cooking results of all sorts of materials such as coal, wax, bones, salts, vegetables, and blood, etc. by heating up a retort that contains the materials, respectively. From each cooking run, Hales collected the *air* (phlogiston) released from the retort by using a uniquely designed instrument and took measurements of its quantity each time (Fig. 3.1). Hales recorded the following findings from coal cooking or pyrolysis experiment (Hales, 1726).

> ...*Half cubick inch, or 158 grains of Newcastle coal yielded 180 cubick* inches of air, which arose very fast from the coal, especially when the yellowish fumes ascended. The weight of this air is 51 grains, which is nearly 1/3 of the weight of the coals.

Although Hales did not further pursue what the *air* was, which is impossible at the time, his work seems the earliest record of a coal gas making in a closed lab setup. Of the ingenuity, however, is the system that he contrived for collecting the *air*, called pneumatic trough, as an effective tool to isolate the target *air* or *airs* from the interference of an atmospheric air or a common air at the time. Decades later, the pneumatic trough was used by a few curious minds such as Joseph Priestley and Henry Cavendish to carry out more discoveries that eventually opened up the Pandora's Box of what air is chemically. About 60 years later, the discoveries that the pneumatic chemistry was about to make would demystify what a fire or combustion is that further led to the establishment of

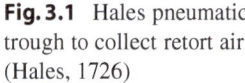

Fig. 3.1 Hales pneumatic trough to collect retort air (Hales, 1726)

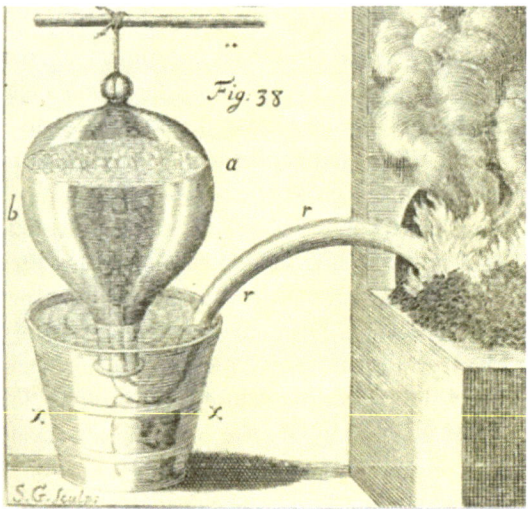

modern chemistry including what's behind the coal gas and gasification. Until then, the phlogiston theory was the most prevailing explanation of a fire or combustion, which is a physical phenomenon.

History indicates that what we know today about coal gas took a century's time to discover and identify each of the associated gases including H_2, CO, CO_2, CH_4, N_2, O_2, H_2S, and NH_3. Some of the organic compounds like COS, C2-C4 hydrocarbons took much longer. Had Hales set a match to the air (coal gas) evolving from retorting the Newcastle coal, he might have found out the *air* inflammable. Then with more *airs* discovered in the following decades, *air* had become more complicated and confusing. To a large part, interestingly enough, it is water and the comprising two elements that held the golden key to unlock the Pandora's Box.

In 1754, Joseph Black (1728–1799), a Scottish chemist, discovered during his Ph. D. study an *air* fundamentally different from the atmospheric air but similar to the *air* released from heating limestone; the *air* of his discovery would extinguish a flame, similar to the *air* released when he reacted magnesia alba (carbonate) with vitriolic acid (sulfuric acid). He named it the *fixed air*. It is the discoveries that made by Henry Cavendish (1731–1810) and Joseph Priestley (1733–1804) provided the decisive evidence that were critical to the birth of the modern chemistry. Both Cavendish and Priestley, members of the Royal Society, were passionate about chemistry and firm believers of the phlogiston theory. Cavendish improved Hales' pneumatic trough by using mercury instead of water as a seal which allowed him to capture other gases, which would otherwise be lost in the water. In 1766, he observed a light inflammable air produced when dropping a piece of iron into a dilute solution of vitriolic acid. He named it *inflammable air*, hydrogen in today's term. Having further improved Cavendish's device and made it multifunctional while sophisticated enough to carry out more complex experiments with gas manipulation

Priestley in 1774 discovered another element that makes water. When heating up mercury calx (oxide) in a sealed flask Priestley found that an air was produced. The air would "burn" amber back into flame when exposed to; and when breathing in the air he felt pleasantly refreshing. Priestley gave it a name of a ***dephlogisticated air***, which is oxygen known later. In the following years, Priestley also discovered more gases such as an inflammable air of carbonic oxide (carbon monoxide), ***alkaline air*** (ammonia), ***marine acid*** (hydrogen chloride) and so on. As time went by, around the year of 1781, Priestley communicated with Cavendish on his another finding of water formation when his ***dephlogisticated air*** and Cavendish's ***inflammable air*** were sparked electrostatically. Becoming interested, Cavendish repeated the experiment in a qualitative way and reported his finding to the Royal Society in 1784 that about 20% of the common air disappeared when reacting with his ***inflammable air*** in a certain proportion. The 20% common air that disappeared should be oxygen in the air. In the meantime, there were drops of dew formed and the dew appeared no difference than the regular water (Brock, 1993).

As far as coal gas or syngas is concerned the following table summarizes the times of gas discoveries and chemists who discovered them (Table 3.1). A note is that historians also credited the Swedish chemist Carl Sheele who independently identified oxygen earlier than Priestley but did not publish his finding until 1777. With regard to the discovery of carbon monoxide there are several accounts. The one with some details is the discovery made by the French chemist Joseph Marie-Francois de Lassone (1717–1788); he prepared carbon monoxide in 1776 by heating zinc oxide with coke but mistakenly regarded it as hydrogen because it gives off a blue flame upon burning (Qiao, 2017). Then in 1800, the English chemist William Cruikshank (1745–1800) identified a compound containing carbon and oxygen. The compound has proven to possess some unique chemical characteristics, which was later found many interesting applications such as in metallurgy, chemical synthesis and catalyst preparation etc. in addition to its value as a fuel gas. Other hydrocarbon gases such as ethane, ethylene and propane etc., typically present in small quantities in coal gas, would have to wait for a few more decades until more had become known about the coal tars and oils that are released during a coal gas making and that would open up another branch of chemistry, the organic chemistry. By looking at the old names of the gases listed here, however, it is quite confusing the least to tell one from the other. Facing the mounting dilemma the phlogiston theory had become challenged.

For whatever the reason, neither Priestley nor Cavendish pursued further to challenge the phlogiston theory with their discoveries that air and water were not elements and the dephlogisticated air (oxygen) would be fixed with the inflammable air (hydrogen), when subjected to a spark, to form water. Understanding both of them were diehard phlogistonists at the time, however, there seems no surprise either. Actually at a point of time, Cavendish believed the light inflammable air of his discovery, hydrogen, was the true phlogiston. What was unfortunate, however, was that England's lead in pneumatic chemistry simply stalled at these discoveries and missed the opportunity to move the

Table 3.1 Discoveries of major gases making up syngas

Current name	Discovered by	Year	Old name
Oxygen	J. Priestley, British chemist	1774	Dephlogisticated air
Hydrogen	H. Cavendish, British chemist	1776	Inflammable air
Carbon monoxide	J. Marie-Francois de Lassone, French chemist	1776	Inflammable air
Carbon dioxide	J. Black, Scottish chemist	1754	Fixed air
Nitrogen	D. Rutherford	1772	Noxious air
Methane	Alessandro Volta, Italian chemist	1776–78	Marsh air
Hydrogen sulfide	B. Ramazzini, Italian chemist	1713	Sulfuretted hydrogen
Ammonia	J. Priestley/Berthollet	1774	Alkaline air
Carbon	A. Lavoisier, French chemist	1789	Carbon

pneumatic chemistry forward to another level. In this aspect, it was the French Chemist, Antoine Lavoisier, who developed on top of these discoveries a platform that had put chemistry on the right track, kicking off the scientific revolution which in turn boosted the industrial revolution, the coal gas lighting and coal gasification.

In retrospect, it takes time to develop and establish a specific field of knowledge and skills to advance the gasification technology as an example; many times back then, however, it took a leap of faith for many great minds to face what their discoveries had actually presented to them. At the time when Henry Cavendish believed that he had discovered the true phlogiston, an invisible, weightless and *inflammable air* that was released when he dropped a piece of iron into a dilute solution of sulfuric acid, but only to prove that he missed the opportunity to identify it as hydrogen, one of the most abundant elements on earth and one of the highly sought green energy as of today. In the delicate experiment performed by Joseph Priestley to heat up mercury calx (mercury oxide) in a sealed flask he found that a gas was produced and furthermore the gas would "burn" an amber back into flame and felt pleasantly refreshing when breathing in the air released, but unfortunately, he named the air as the *dephlogisticated air* instead of oxygen as we know of today. Equally unfortunate is that both of them missed the precious opportunity to recognize the nature of water in their experiments where water drops formed when the *inflammable air* gets burnt in air or in the pure *dephlogisticated air* though they might have been puzzled at that time by the fact that the water drops formed on the wall of the flask is no difference than the water present everywhere else. Although the credits for discovering hydrogen and oxygen are still given to them respectively, it is another great mind, the French clergyman Antoine Lavoisier who named hydrogen (water-maker) and oxygen (acid maker) that we are using today. Lavoisier then moved on further to challenge the legitimacy of the Aristotelian principle in a systematic way that eventually led to the establishment of our modern chemistry. It also laid the foundation for gasification starting to exist.

The Chemistry of Gases

In France, the young Antoine Lavoisier (1743–1794), a civil servant and a passionate chemist in his spare time, was born into a family of lawyer in 1743. At age 20, Lavoisier obtained a law degree from the College of Mazarin. While in school Lavoisier became interested in chemistry and took chemistry classes offered by Guillaume-Francois Rouelle (1703–1770) who established his new theory of salts along the line of the phlogiston theory. Rouelle classified salts based on their crystalline forms, the acids and bases from which the salts were prepared. This exposure had impacted Lavoisier of his future thinking toward chemistry. Later during 1763 and 1767, Lavoisier studied geology under one of his family friends, Jean-Etienne Guettard (1715–1786). As an assistant to Guettard they performed a survey of mineral resources across the open field of France. It was during this time that Lavoisier also paid attention to water purification and related chemistry. He studied closely gypsum and its crystalline forms, a popular material painted walls around Paris, and presented a paper about his finding to the French Academy of Sciences in 1765, the prestigious institution that was set up by King Louis XIV in 1666 to encourage and protect the spirit of the French scientific research and to provide services to address any scientific questions officially raised to the body. Being ambitious to become a member of the institution Lavoisier in 1764 participated in a prize contest for lighting up the Parisian streets, which won him a special medal from King Louis XV two years later. The prize contest was awarded by the French Academy of Sciences as part of the French government's incentive policy to encourage innovation. Earlier during the period of the Seven Year War (1756–1763), the French government realized the Britain's significant lead over its industrial strength aided in large by the advancement of science and technology and, therefore, reassessed its own policy as a conscious decision to address its urgent strategic needs. Another example stimulated by the program was the Leblanc Process invented by Nicholas Leblanc (1742–1806) in 1789, the start of the early chemical process to make soda. Soda was an important ingredient that was used for many purposes other than baking, for examples for making soap, paper making and glass making and so on. Before the Leblanc Process, soda had previously been made from natural plants or natural minerals. Back to Lavoisier, the talented young man was finally selected as an assistant chemist, a ranking member of the Academy of Sciences in 1769. Using the inherited wealth from his family in the same year Lavoisier invested in the Ferme Generale, a private company contracted to the government for taxes collection on Tabaco, salts and imported goods.

This turned out to be a bad investment that ended his life too soon during the French Revolution.

Being passionate about chemistry Lavoisier had devoted all his spare time to science, studying chemistry and experimenting with reactions to clarify or confirm doubts or questions that were raised by the public or that he might encounter. Over time he spent almost all his fortune purchasing and making the most expensive equipment and building the best laboratory system in Europe at the time so to provide him with the necessary functions and accuracy for a complete quantitative analysis, capturing what's in and what's out. Although knowing well the phlogiston theory, Lavoisier believed and took an approach that mass must be conserved before and after any change or transmutation. The more experiments he performed the more confusing he was becoming about the phlogiston theory. He picked up the word *gas* first coined by Black in 1756 and further proposed that there must be a gaseous state existent; that is, all matters should have three states, solid, liquid and gas depending on the conditions they were at. Using water as an example, it exists in a solid state when the temperature is below the freezing point, a liquid state at room temperature, and a vapor state when heated up beyond the boiling point. Such thinking, out of the box, had opened up his mind to do more about air and combustion. In his experiments of burning sulfur and phosphorous, he found that both these elements gained weight, respectively, instead of losing it by releasing phlogiston. Knowing similar results reported with metal when roasted in air and the formed calx was stable and could only be reduced by burning in charcoal, he believed that the atmospheric air somehow got involved in the reaction and became fixed in metals causing the weights to increase. More experiments produced more conflicting results, making Lavoisier start to become suspicious of the legitimacy and unpredictable nature of the phlogiston theory. He believed that the phlogiston should be predictable and consistent by having either positive weight, no weight or negative weight, but rather not being anything just for the sake of explaining any specific encounters. Lavoisier knew that he would need more data to support his suspicion. Having realized many similar works including the work on pneumatic chemistry happened in England Lavoisier spent the whole year of 1773 studying the earlier works carried out by others. What he logged into his lab notebook on Feb 20, 1773 (Brock, 1993) clearly indicates his intent on why he wanted to repeat the works performed by others.

> I have felt bound to look upon all that has been done before me as merely suggestive. I have proposed to repeat it all with new safeguards, in order to link our knowledge of the air that goes into combination or is liberated from substance, with other acquired knowledge, and to form a theory.

At times, he thought that combustion or firing was the reaction between carbon and fixed air (carbonic acid or carbon dioxide). So did Priestley who thought of his dephlogisticated air the nitrous oxide when he prepared the mercury calx from mercury nitrate. When Lavoisier demonstrated to the public the high profile experiment by using a big lens under the sun as heat source in 1789 the fact that both diamond and charcoal give off the same carbonic acid in a closed atmospheric air under the intensive heat of the big lens is a simple proof that both are made of carbon. What Lavoisier did differently from others is that he took measurements of what's in and what's out, allowing him to grasp the total picture.

What made Lavoisier pull the trigger on the phlogiston theory, however to a certain extent, are the discoveries made by Cavendish and Priestley between 1774 and 1783 about the light inflammable air and the dephlogisticated air. That provided Lavoisier with the decisive evidence that helped him decipher what water and combustion are, break up with the phlogiston theory, and finally start to build a completely new system toward modern chemistry. In late 1774, the year that he made his discovery of the dephlogisticated air, Priestley traveled to Paris and met with Lavoisier there. Priestley informed Lavoisier of his finding and experience with the dephlogisticated air. Lavoisier repeated Priestley's experiment with mercury and its calx in his closed experiment system. The reproduced results made him think that it was this pure air that increased the weight of metals and non-metal elements, which the latter produced acids when dissolved in water. Eventually, Lavoisier was able to close the gap by the fact that the pure air was part of the regular air that humans breathe in, implying that air was not an element. Five years later in 1779, he named the dephlogisticated air or pure air oxygen, a Greek word meaning acid maker, as it reacted with non-metal elements producing acids. At this time what held Lavoisier back was the fact that the reaction between a metal and an acid released an inflammable air; but the calx of the same metal had to be reduced by burning together with charcoal to return back to metal form. This had puzzled him until nine years later when Cavendish's secretary, Charles Blagden visited him in Paris in 1783. Charles shared with Lavoisier of what Cavendish had done recently on a water formation between the light inflammable air and oxygen (the dephlogisticated air). This inspired Lavoisier that a water was not an element either and prompted him finally to firm up his combustion theory that it was oxygen exploded with an inflammable air to form water; and it is the same oxygen reacting with metal to give a calx with more weight; the inflammable air released from the reaction between a metal and an acid was not from the metal but the water where an acid is dissolved in. He gave the name of this light and inflammable air as hydrogen, water maker in Greek. By 1785, Lavoisier seemed getting ready to abandon the phlogiston theory which he believed its mysterious and unpredictable nature non-scientific. To complete his new theory, the theory of oxygen combustion, however, there was one more piece that he would need to make it right, that is where to place the element that caused heat and light, instead of phlogiston, during combustion. He coined a word, Caloric, present in oxygen not in metals containing heat and light to be released when burned. Lavoisier presented new findings in his famous treatise"**The Traité élémentaire de chimie (Elementary Treatise on Chemistry)**" in 1789.

Based on the oxygen theory of combustion proposed by Lavoisier, matters (metal and non-metal) react with oxygen (containing caloric replacing the word of phlogiston) in atmospheric air to give off calces, resulting in a weight increase, while the caloric released to produce heat and light. Lavoisier's new theory reasonably explained the combustion process except for leaving the caloric still in the air which needs to be corrected about a half century later. Lavoisier's combustion theory, though not perfect, had removed the mysterious veil with regard to flame or fire and put chemistry onto the right track, kicking off the scientific revolution in the coming centuries. It certainly helped ease public fears when coal gas for lighting would come to general public lives in the next century, by explaining what coal fire was and why it burnt.

In conclusion, no one would have believed that it was the two natural materials, air and water, being so familiar and vital to the existence of human beings and their living ecosystem, also held the golden key back then to open up the Pandora's Box about the essence of fire or combustion. Had the two materials been deciphered earlier, one may wonder, would human beings be able to get out of the Renaissance earlier and to start off the scientific revolution earlier? Well, there seems no way to answer them but one thing is clear, deciphering air and water and knowing their chemistry had certainly cleared up the way for the coal gasification to come to existence as they are two critical reacting media and agents to convert hydrocarbons including coals and wood chips, biomass and other municipal wastes etc. into inflammable gases which in turn made them indispensable utilities for public lighting, heating, and many more uses with industrial and metallurgical processes during the industrial revolution that was about to come.

References

Brock, W. H. (1993). *The Norton History of Chemistry*. W.W. Norton & Company Inc.

Hales, S. (1726). *Vegetable Staticks*. London.

Qiao, L., & N. Z. (2017, Jan-Mar). Carbon monoxide as a promising molecule to promote nerve regulation after traumatic brain injury. *Med Gas Research*, pp. 45–47.

The Era of Gas Lighting

4

Prior to the coal gas lighting, the world as a whole used animal fat, whale oil and vegetable oil to make candles to obtain illumination. Most likely, Lavoisier's 1764 contest of the street lighting was based on candles or oil lamps in his design. Day time in the northern hemisphere such as in England is relatively short, especially during the winter season. Stretching an hour or two into a day after sunset would certainly be attractive for mills and factory owners to boost their profits or simply to get more things done. Both candles and oil lamps, however, were very limited in lighting power noting that one candle light is equivalent to 12.5 lumens or one watt of incandescent light; and neither were they cheap because of the limited supply of materials, which were worsened from the competing demand for soap making back then. For a typical cotton mill operation in Britain, therefore, such a cost on annual basis could go easily above thousand pounds just to claim 2 more hours of time from a day. For the general publics except those wealthy few, candles and oil lamps had remained a sensational luxury to enjoy. Any economic alternatives to candle lighting would be attractive and highly sought after, which had become a pressing issue by the end of the eighteenth century with towns and cities rapidly expanding, especially in England because of its extensive industrial activities.

The Illuminating Gas

When coal gas lighting arrived at England at the beginning of the nineteenth century, however, it was first met with different reactions from the public. Some embraced it with excitement and others reacted with fear and suspicions due to the lack of knowledge about what the first of the kind is. It eventually found its niche plays in areas such as street

Q. Zhuang, *From Coal to Hydrogen*, Synthesis Lectures on Chemical Engineering and Biochemical Engineering, https://doi.org/10.1007/978-3-031-55586-2_4

lighting urgently in search for economical alternatives, which was a significant event that had changed the public life forever. Like the Westminster Review, a Victorian periodical wrote in the early years of street lighting with coal gas (Long, 2022).

> The introduction of gas lamps would do more to eliminate immorality and criminality on the streets than any number of church sermons.

Along with the progress made with pneumatic chemistry, the combustion principle, and its chemistry the mysterious veil over combustion or fire had been gradually peeled off. Like what Hales did with his retort experiments to measure the quantities of "air" produced from different materials, many philosophers, chemists or engineers quickly started to look into other materials for a competitive source of inflammable gases for lighting purposes. Cracking oils out of either animal fats or resins, cooking wood biomass in a closed pot, and making hydrogen from the recent discoveries by Cavendish or Felice Fontana (1730–1805), an Italian physicist, are some of the trials made at the time. These sources were all unlikely feasible due to their limited supply and high prices, similarly under the spell of candles and oil lamps.

In 1799, Philippe LeBon (1767–1804), a French civil engineer by training, filed a patent in 1799 for a device that he called the Thermolamp that would make an inflammable gas from the distillation of wood for lighting. Later in 1801, LeBon displayed his Thermolamp in Paris in order to attract financing to start a business but failed to do so. It is the Scottish engineer William Murdock who developed this new way of lighting into a package, a process and equipment to manufacture coal gas by retorting coal in 1792 that had forever changed the lives of modern society.

William Murdock (1754–1839), a Scottish engineer at the Boulton & Watt, became interested in the inflammable gases during his assignment to Redruth, Cornwall for steam engine installations at coal mines during 1792–1798. Around 1792 he lighted up rooms of his house, currently a landmark, in Redruth with coal gas that he made by cooking coal in a simple iron pot as a retort next to his house; and between the retort and his house, he used a tinned iron and copper tubes of 70 feet long to guide the coal gas to his house (Hughes, 1853). In the following years, he continued the work in his spare time with a local company to further develop the retort, investigated gases produced from different coals and necessary components such as coal gas cleaning up, storage, distribution, and lighting device etc. What he had done during this period laid a solid ground for later deployment.

Having finished his assignment and returned to Birmingham where the Boulton & Watt Soho Works was located, Murdock built a small scale retort system to produce coal gas and used the gas to light up the main building of the Soho Works. After both Boulton and Watt retired, their sons, Matthew Robison Boulton and James Watt Jr. took over the Soho Works business and became interested in the coal gas business once becoming aware of the LeBon's undertaking in Paris in 1801. Taking the opportunity the next year to celebrate the Peace of the Amiens, a 14 month temporary treaty to end the hostility between France and Britain, Murdock was asked to set up his retort lighting at the Soho Works building. On March 25, 1802, Murdock placed a small retort at each end of the Soho Works building and lighted up the coal gas coming out of the retort and connected a short distance above the retort to a burner. The unique neo-light astonished passersby and added more colors to the celebration. This event had become the first public lighting in a practical way, showcasing the new technology that signaled the arrival of a new era of lighting at the early stage of the industrial revolution. With the coal gas lighting technology becoming established in England, coal gas lighting soon spread to both sides of the British Isles, the continental Europe and the United States of America, ushering the coal gas making technology and the industrial revolution to another level.

Coal Gas and Public Utility

Having been selling steam engines and related services to mills and mines in the industrial sector since the 1770s the Boulton & Watt Co. had developed quite a customer base which would naturally become the target customers for Murdock to develop for his new coal gas lighting business. This strategy worked well for a quick start. At the beginning of 1805, Murdock assisted by Samuel Clegg (1781–1861), an apprentice at the time but later becoming a key and independent individual in delivering the coal gas business, commissioned two coal retort projects in parallel the following year, one at the Messrs. Phillips & Lee's cotton mill at Salford, Manchester and the other at Mr. Henry Lodge's cotton mill at the Sowerby Bridge, near Halifax. Being one of the largest cotton mills in Britain, the Phillips & Lee's mill was benefited from the deployment of new technologies of textile related, steam engines and associated equipment. The General Manager Mr. Lee immediately became interested in coal gas lighting, but as a precaution, executed the coal gas making and lighting in stages by first lighting up one counting room and Mr. Lee's dwelling nearby. The project of coal gas making, distribution and lighting up the rooms went smoothly in general. What was unexpected, however, was an odor and smell from the coal gas burning that had become intolerably irritating, especially when doors and windows were left closed. It is caused by the sulfuretted hydrogen (hydrogen sulfide), known for a long time, present in the coal gas. Then improvement was followed by adding lime washing in the gasometer or gasholder, a device to store coal gas before being distributed for consumption, to remove the sulfureted hydrogen to a tolerable level. By this change the project went on by distributing coal gas to all other rooms at the mill via pipes totaling seven miles long. The mill installed 271 Argand burners and 633 cockspur burners in all rooms for lighting. The cockspur burner was designed by Murdock at the mill (Samuel Clegg, 1841). During winter time, coal gas ran continuously, a successful event that made a milestone to kick off the age of gas lighting. In February 1808, Murdock presented his project demonstrated at the Phillips & Lee's mill to the Royal Society,

"*An Account of the Application of the Gas from Coal to economical Purposes*", showcasing a very attractive economics as summarized below (Murdock, 1808).

On annual average, extending two hours a day would require 2,500 candles, costing about 2,000 pounds. With coal gas, 2,500 cubic feet of coal gas would be necessary to load all 904 burners, producing equivalent light power that 2,500 candles would provide. To manufacture 2,500 cubic feet of coal gas daily or 782,500 cubic feet annual required 110 tons of cannel coal as feedstock to for retort and 40 tons of common coal as fuel. Coal gas production would result in 70 tons of coke for sale; tar and others had no value. Considering interest payment on capital and tear & wear to equipment, the annual expense would be at 600 pounds, indicating a significant saving to owner comparing to the 2,000 pound cost for candles.

What a significant saving that the coal gas lighting could bring to the Messrs. Phillip & Lee's cotton mill. Convincingly, coal gas lighting had demonstrated its competitiveness from the very beginning.

While Murdock's account on profit comparison might be of a bit aggressive or a best case scenario from a long term sustainable basis as later experience would prove more improvements and modifications necessary to smooth out the coal gas making and lighting system. As the process became complex by adding more units and devices such as coal gas purifier for the removal of hydrogen sulfide, equipment for the recovery of coal tar, ammoniacal liquor and light oil, gas meters, pressure regulators, etc. Both capital expenditure and the costs to operate and maintain would add up to the total. Nevertheless, coal gas for lighting proved itself clearly a viable and competitive alternative to replacing the old and dim candle light or oil light. When Leblanc died in 1806, the Leblanc's process for soda making had not achieved success. Murdock had, however, demonstrated the viability of coal gas lighting by putting his investigative results and know-how that he had acquired in the past sixteen years all together into a working coal gas lighting product, a process system, which also marked the beginning of the practice of chemical process engineering in its primitive form. In the following decades these two industries or , in a more precise term, these two chemical processes would cross each other several times, significantly impacting and advancing each other's development.

While Murdock at Boulton & Watt was trying to develop more coal gas works in the industrial sector across England, Frederick Winsor (1763–1830), a German inventor and entrepreneur, was working diligently as well to explore another market for coal gas in England. It is the market of public utility for lighting streets, commercial buildings and residential houses. Being also exposed to the LeBon's Thermolamp during his visit to Paris in 1801 Winsor believed that coal gas should play a big role in the public sector to light up streets, hotels, theaters, bars and so on to improve the public life as a result of the rapid economic development during the early industrial revolution. His business model seemed straight forward; that is, to build a large central coal gas works to manufacture coal gas and a pipeline system to distribute the coal gas to light up public streets, bridges, hotels, theaters and houses while selling by-product coke to the market as a clean fuel for cooking, heating, and other domestic uses. This unique business model, however, was completely new and received lots of doubts and critics in addition to fears knowing that the chemistry of coal gas firing at the time had not yet reached far into the public. To demonstrate his business concept Winsor put up a showcase of several street lights fired with coal gas standing conspicuously at Pall Mall in 1807. Such a showcase as an effective marketing approach did attract a significant amount of public viewings accompanied by excitement, curiosity and fear on those faces as depicted in the famous sketch by George Rowlandson in 1809 (Fig. 4.1). In the meantime, Winsor had been seeking interested investors to raise funds while lobbying the Parliament for a charter to start his business. It had taken Winsor some extra time but his perseverance was paid off; in 1810, the Parliament granted Winsor a charter to develop coal gas works to supply coal gas to the Westminster and adjacent suburbs for public use. Once raised enough money Winsor set up a company, the Chartered Gas Light and Coke Company ("GLCC"), in 1812 and put it right into work. However, the road to coal gas lighting seems not a smooth one, at best. Having suffered many initial setbacks of both technical and commercial, GLCC eventually emerged as the pioneer in developing and operating the coal gas business for lighting and had become one of the largest coal gas companies in England. GLCC's first coal gas was manufactured from the gasworks built on Peter Street, Westminster, which was commissioned in September 1813. Three months later on New Year's Eve, the coal gas lighting replaced the oil lamps on the Westminster Bridge. A year later on Christmas Eve of 1814, coal gas lanterns also substituted the old oil lamps at the St. Margaret's parish, beginning the general lighting of coal gas in London (Guide G., Gas Light and Coke Co.).

A hundred thirty-seven years later, after WWII in 1949, the GLCC became nationalized and merged into the North Thames Gas Board, one of the twelve state owned gas boards cross the country. In 1986 it reemerged as the British Gas Corporation out of the privatization efforts, the Gas Act 1986.

Fig. 4.1 A peep at the gas-lights in Pall Mall (Rowlandson, 1809)

A PEEP at the GAS LIGHTS in PALL-MALL.

The Go to Market Strategy

Although the coal gas for public lighting business had become prospered, Murdock's coal gas business had remained almost stagnant. Boulton & Watt eventually withdrew its efforts from the coal gas making business. Why? In hindsight, the competition with Winsor might have contributed to their withdrawal. Then, Murdock's inaction to obtain necessary patents for his development of the coal gas making process could also bear the blame. From a commercial perspective, Boulton & Watt's strategy to focus on its existing customers in the textile sector did provide them with a quick start. It was, nonetheless, a relatively small niche and captive market where business owners were routinely conscious of any capital spending due to the fierce competition at the time. Although coal gas works built into mills are generally small in scale it takes similar efforts to develop. After all, Boulton & Watt's business was equipment manufacturing and installation services, different from the process driven coal gas making business that required other scientific disciplines than the manufacture of equipment like steam engines.

Technically speaking, the early success that GLCC had achieved would have hardly been possible without two individuals, Samuel Clegg and Frederick Accum (1769–1838). It was their collective knowledge of chemistry and extensive hands-on experience that played vital roles in rendering Winsor his business plan well executed to deliver the coal gas for lighting. Frederick Accum (1769–1838), a German chemist, practiced his chemical laboratory and equipment business during his living in London between 1793 and 1821, such as by selling and making standard reagents and equipment. During 1801 and 1803 he worked as a laboratory assistant to Humphry Davy (1778–1829), an English chemist and inventor, at the Royal Institution and then pioneered public lecturing on practical chemistry in Britain, which made him a prominent figure in the subject matters like chemistry, gas lighting, and food safety associated with chemical use. He shared Winsor's vision to use coal gas for public lighting and assisted Winsor with experiments about the new gas lighting business. He also helped the Parliament Committee in reviewing Winsor's application for a charter to launch the Gas Light and Coke Company to conduct coal gas lighting business. Because of his influence and knowledge of chemistry, Accum later became a board member of the then GLCC incorporated in 1812. His publications in 1815 and an updated one in 1819 (Accum, 1815, 1819) are the earliest available documentation about coal gas making and related business back then, which serve as good resources for those who are interested in to grasp what it was alike at the beginning of the coal gas making. To build coal gas works and make it work, however, it requires a different discipline, and it was Clegg who possesses an extensive hands-on experience coming to the rescue of Accum's work at the Curtain Road.

With coal gas making gradually improving its process nature would have become more significant, drawing more chemical knowledge and process skills in order to deliver a sound coal gas works. Examples were Clegg's invention of the purifier of using lime to absorb sulfuretted hydrogen, and Accum's knowledge related to safety and effluents like coal tar and ammoniacal liquor that would soon have to be dealt with due to the increased public awareness. To Clegg, Murdoch seems to have lost the cutting edges in terms of necessary know-hows on process and chemistry knowledge, which had become more and more critical in order to establish a coal gas making system that would warrant both a satisfactory operation and a quality coal gas meeting the evolving requirements for the gas lighting applications. Nevertheless, Murdock's pioneering ingenuity of exploiting the value of the coal gas, contriving the coal gas making apparatus and deploying it for commercial purposes would not be discounted at all; he is remembered by delivering a brighter neo-light without using a wick which had significantly impacted every corners of the British society during the ongoing Industrial Revolution, well into the twentieth Century.

The GLCC's early success, however, was not without hardship considering the time when the coal gas lighting was something completely new and first of the kind. General public was not familiar with coal gas and the chemistry involved. Now that the inflammable gas or the spirit of fire would be introduced into their neighborhood, streets and houses, fears out of no reason or fears over superstition and explosion were just part of the natural road blocks, which were time to time worsened by the accidents that took place with the operation of gas works. For example, not long into the operation around the end of 1813, the gasworks at Peter Street experienced an explosion due to an unexpected backflow of coal gas from the purifier for sulfuretted hydrogen removal. The explosion shattered windows of houses in the neighborhood, and severely injured Clegg as well. Then another disaster followed and hit the company hard on the night celebrating the Peace of June 1814. To impress the attending allied dignitaries, the event planner built a pagoda tower of an oriental style and dressed it up with more than 10,000 gas lamps to be lighted up with coal gas supplied from the gasworks at Peter Street. It is supposed to be the centerpiece for the night of the peace celebration. As an extra precaution, a rehearsal was carried out the night before the event. The coal gas lighting surrounding the pagoda was splendid. On the night, however, the rumor is that Sir William Congreve did not follow the procedure. Instead of lighting up the coal gas lamps first, he set off fireworks first. What happened next was disastrous; the sparks from the fireworks triggered an explosion to the coal gas lighting system due to gas leakage. The explosion simply brought down to ground the eighty feet tall octagonal pagoda tower (Major, 2018).

Then, other examples are that oil lamp lighters were afraid of lighting up the gas lamp at the Westminster Bridge at the beginning, and that the gasometer had to be made small and housed in a brick and mortar building for protection in case of explosion, which the latter is obviously contrary to the safety practice nowadays. For some gas works, building multi-gasometers instead of a large one plus the housing structure certainly added up to the total cost of the projects. It, however, reflected public concerns over this emerging chemical process and its dangers or mysteries associated with the inflammable coal gas. These setbacks had indeed created quite some headwinds to move forward the coal gas lighting business. GLCC, however, was able to manage and overcome the initial difficult time and to advance its business by developing more coal gasworks. By 1819 GLCC had placed two more gasworks up into service, one on Curtain Road and the other one on Brick Lane in London. All three gasworks manufactured coal gas and was distributed through a network of 288 miles of pipelines laid underground connecting the gasworks with streets and houses. Such a coal gas lighted up more than 51,000 lamps in the Metropolis area of London alone. By 1823 each of the three gasworks produced about 300,000 cubic feet (8,000 cubic meters) of coal gas daily by retorting 24 chaldrons of coals. A note here is that the chaldron is a unit to measure how much coal going to a retort for coal gas making, not by weight which is more typical. A chaldron is somewhat a standard wooden boxed wagon driven by animals, typically holding about 2,837 lbs. or 1,287 kg of coal. Another frequently used unit for weight is coal weight, cwt. One chaldron is about 25.33 cwt; a cwt is 112 lbs. Therefore, the 24 chaldrons of coals used for coal gas making at each of the gasworks is 31 tons of coals during a normal operating day.

Generally speaking, Winsor's immediate success in coal gas lighting confirmed his vision; that is, the coal gas had responded well to the urgent need of the public lighting that the bustling cities and towns were actively seeking. In the meanwhile, coal gas making itself could deliver a viable and competitive solution to meet such a need as first proved by Murdock. Such a coincidence made the fame for both Murdock and Winsor in terms of their ingenuity and vision so that the coal gas business found its immediate market, entered into a fast lane of growth, and enjoyed a long prosperity for more than a hundred years. Once the streets in London had been lit up with coal gas, a brighter, safe, and neo-light, coal gas lighting would soon spread quickly beyond London into cities such as Edinburgh, Glasgow, Liverpool, Bristol, Bath, Cheltenham, Birmingham, Leeds, Manchester, Exeter, Chester, Macclesfield, Preston, Kidderminster, and many more towns and villages across the country. By 1830 there were 200 gasworks carrying out coal gas business; the number went up to 800 in 1850 and soon reached over 1,000 by 1860 (London). Coal gas making had joined the main stream of the industrial revolution, opening up the age of coal gas that had lasted for almost one and half centuries of time. More interestingly, the coal gas lighting had become a part of London culture and a nostalgic subject of many movies and literatures. During this period, coal gas making business and technology continue to improve, evolve and, at times, have to self-reinvent in order to stay afloat or to adapt to the changing market conditions and regulatory environment.

Coal Gas Lighting Process

Like any other new technology, even in today's environment, the road of commercialization from a small scale to an industrial scale requires time, knowledge, expertise, and resources. Learning from problems and difficulties that arose from actual operations, fingering out the root causes of those problems and difficulties, then identifying ways to address them by improvements and modifications to the corresponding process or equipment are the typical methodology. This seems especially true for coal gas making considering the time when there was so much unknown, chemistry in its infancy and chemical process industry essentially not yet in existence. It would be interesting to look back at how the commercialization process for coal gas making, a process technology in today's term, was alike in its making, starting from scratch to maturity; and during which how Murdock's passion and engineering skillset and Clegg's ingenuity and his exchanges with great minds like Dalton and Faraday etc. played into the development of the coal gas making process. One thing seems clear chemists and engineers were eager to pick up whatever had come out of the scientific discoveries and inventions and to put them into practice and use. Learning by practice is the way of life.

Coal gas making is a straight forward process, starting from the piece of equipment called retort. Murdock designed several types of retort as shown in Fig. 4.2 during the early days. During coal gas making, coal is placed into a retort (**A**) through an opening and sealed the opening properly afterward. The retort is then heated up by a coal fireplace (**C**), typically using cheaper coal than the coal used in the retort, to a temperature up to around 1,000 °C. Subjecting to such a high heat, like what Hales observed in his study back in the 1720s, coal in the retort goes through a series of chemical changes to release coal gases, which would continue for a few hours and then slow down. The chemical changes were totally unknown then and are still so to some extent even today. The coal gases released flow via exit (**B**) and are collected in a storage tank, large enough to hold all the coal gases produced from the whole cycle operation, before being used for lighting or other purposes. In principle, the process could be as simple as this, just like what was done with the early gas works at Mr. Lee's cotton mill and Mr. Henry's Lodge. Murdock did a decent job in engineering the necessary hardware such as retort, fireplace, gasometer, piping, cocks, and burners etc. to produce, store, and transfer coal gas for lighting. The operation runs in a batch mode; each cycle takes about 8–12 h depending on the temperature and the coal in the retort. Upon finishing one cycle, the remnant coke is removed from the retort, then reload with fresh coal, and another cycle starts. It was a manually operated process, starting from the early morning and storing the manufactured gas in gasometer for lighting when the sun sets in.

In reality, however, there is another side about the product, the coal gas that has to be considered, customer satisfaction. Customers at the early installations complained about the smell and couldn't stand for it especially in a closed room for example during the winter time for an extended period of time. A new challenge arises from the sulfuretted hydrogen and ammonia etc. inherent in the coal gas produced, which requires the knowledge of chemistry to address. This seems the point where applied chemistry started to interact with, not yet the retorting itself, but the resulting coal gas by finding ways to treat and process the impurities such as sulfuretted hydrogen and ammonia and later on the coal tars etc. Measures to the problems did not get resolved at once but had taken place over time in correspondence to forces driven by market conditions, commercial

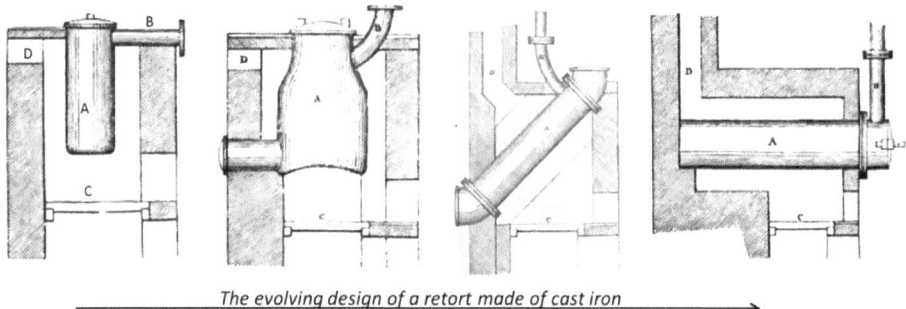

The evolving design of a retort made of cast iron

Fig. 4.2 Early cast iron retort (A: Retort; B: Coal gas outlet; C: Furnace; D: flue gas exit) (Samuel Clegg, 1841)

and regulatory environments, and available technology. This is where Clegg's chemistry background comes to play.

While young, Clegg received an education in chemistry tutored by Dalton at the Manchester Academy. Dalton's teaching certainly helped Clegg's future achievement in developing coal gas making system including the purification using lime to remove sulfuretted hydrogen (hydrogen sulfide), which turns out to be an inherent component of coal gas making. After joining the Boulton & Watt Co. in Birmingham, Clegg assisted Murdock in developing and constructing coal gas works. During which, he had become interested in doing it. Samuel Clegg, Jr., Clegg's son, described in 1841 in his publication *"the Practical Treatise on Manufacture and Distribution of Coal Gas"*, that Clegg had benefited enormously from this period of experience.

Left the Boulton & Watt about 1805, Clegg made his name by building half a dozen coal gas works for mills, shops, and institutions such as the Harris of Coventry in 1807, the Catholic College of the Stonyhurst, Lancashire around 1808, Mr. Greenway of Manchester in 1810, Mr. Samuel Ashton and Brothers at Hyde near Stockport, and Mr. Ackman's printing shop in London in 1812. Through these projects, Clegg was able to implement Murdock's coal gas making system from a primitive gas making system to relatively a fully functional process system including coal gas making, gas cleaning, tar collecting, metering, and storing etc., which produces and delivers a satisfactory coal gas for lighting purpose. Upon joining the GLCC in 1813, Clegg was tasked to review the print of the gas works already under construction at the Curtain Road. Clegg's review suggested that the print was inadequate and the ongoing construction of the gas works, if built, would hardly work. Listening to Clegg's suggestion GLCC decided to open up a second site on Peter Street, Westminster and put Clegg in charge of it. Clegg did not let the company down and, in the same year, successfully delivered about 14,000 cubic feet of coal gas daily for lighting up about 4,000 lamps including those on the Westminster Bridge and in St. Margaret's parish.

According to Clegg Jr.'s 1841 treatise and Murdock's, 1808 presentation to the Royal Society, the early coal gas works built by both Murdock and Clegg for cotton mills at either the Phillips & Lee's mill or Mr. Henry's Lodge near Halifax proved successful in terms of coal gas making and for lighting, by any standard though being primitive. From a process flow perspective, using the gas works at Mr. Lee's cotton mills as an example, coal is loaded into a mesh bucket or cage and then placed into a vertical retort; underneath is a fireplace to heat up the retort (Fig. 4.2). Upon heating, coal gas evolves and leaves the retort hot, then cools down on its way to a gasometer (storage tank), and cooling continues. What is happening during the cooling is that tar, light oil, and water vapor in coal gas condense and separate, one after another, from the gas phase; the lower the temperature it goes the more condensation takes place. Along with the condensation of water vapor component of ammonia becomes dissolved in water to form ammonia liquor, somehow mixed with other liquids inside the pipe and gasometer. At this time, there are three pieces of major equipment involved in coal gas making, furnace, retort and gasometer. The most left retort design in Fig. 4.2 illustrates what was adopted at Mr. Lee's mill, in which the retort is integrated into a fireplace right beneath. Two cast iron retorts in operation handling about 300 lbs. of coal for coal gas making. To start coal gas making, first, load the cage with coal and then lower the cage with a crane into the fireplace through the opening at the top; then start the fireplace to heat up the retorts, which typically takes some time. The fireplace uses the same coal or a cheaper one, whenever available, as fuel consumes about 30% of the coal use in retort. With the temperature going up coal gas starts to evolve from the retort; the coal gas was conveyed by iron pipes into a large reservoir or gasometer, where it is washed, purified, and stored before being conveyed or distributed through pipes, called mains, to the mill for lighting. At the end of each operation lift the cage out to dump out the residual coke; then start another cycle. This description pictures the earliest coal gas making process for lighting application, in its simple and primitive form as a starting point. Obviously, Mr. Lee's complaint about the smell tells that Murdock's design by using water to clean up the coal gas was inadequate. As noted in the previous section, the sulfuretted hydrogen was noticed by the Italian physicist Bernardino Ramazzini in 1713 as inflammatory and causing asphyxia; it is a gas typically found in city sewers; so is ammonia. The French prolific writer Victor-Marie Hugo described the sewers in 1800s' Paris as "*the intestine of the Leviathan*" (Smith). Lime had been known for long as an effective absorbent to remove the smell. Clegg made use of it as well but took him a few projects to try out different designs and ways to apply lime before finally landing on a reasonable solution that solved the problem. Process engineering emerged.

To reduce the sulfuretted hydrogen in coal gas in the next project at Mr. Harris of Coventry, just southeast of Birmingham, Clegg placed a layer of lime at the bottom of the gasometer to absorb hydrogen sulfide from the coal gas. It worked, though not ideal; the smell of the delivered coal gas upon lighting reduced to a tolerable level. Soon a cumbersome surfaced. The spent lime needs a regular replacement; it is a difficult and tedious job to replenish the lime in and out of the gasometer. Around the same time of 1807–1808, Clegg worked on another gasworks for the Catholic College of Stonyhurst in Lancashire, the first college to be lighted up with a coal gas. Clegg was able to carry out some experiments of mixing the lime with water by collaborating with professors of the college. Based on the results obtained Clegg designed a separate vessel, a tank, upstream of the gasometer to hold the lime water to purify the hydrogen sulfide in a coal gas. The stand-alone tank was equipped with an agitator to stir the lime water, when necessary, to facilitate a good contact of lime with coal gas, which comes in from the bottom of the tank. Such an improvement achieved a better result than his previous design and purified the coal gas satisfactorily for lighting, which had been widely adopted in other projects around that time. The updated design was then applied to several more coal gasworks he developed in the next few years. With more gasworks up running issues and problems around a coal gas making get exposed one after another. The next problem is the tars formed from the coal gas making easily plugged the piping when cooling down on its way to the gasometer, which is getting worse when mixing with dust carried in the coal gas and the piping needs frequent cleaning; then the tars as a waste has to be disposed of, often ending by piling up in an open area, which eventually becomes another problem to the surrounding neighbors. Around 1810, when building a coal gasworks at a cotton mill with Mr. Greenaway of Manchester, Clegg installed a hydraulic main of his invention. It is a large bore cylindrical vessel laid horizontal above the retort and connected to the rising pipe for coal gas (as an example from Fig. 4.3, D). Once becoming cooled down in the hydraulic main the tars would condense, separate from the coal gas and flow into a tar cistern for storage. The coal gas, now almost free from coal tars, leaves the hydraulic main to downstream. This was an important implementation and had been used during the era of coal gas making. In the next few jobs, Clegg made the lime water purifier and the hydraulic main standard features for gasworks at the time including the gasworks at Peter Street that Clegg managed after joining the GLCC in 1813. The business model of the GLCC, however, was different from all the owners that Clegg had worked with in the past. The coal gas with previous owners was for a captive use, basically under one owner who would foot the bill for the whole coal gas making operation. For GLCC, however, the coal gas would be distributed to different jurisdictions, districts, and customers spreading out many miles away from gasworks. In 1815, Clegg invented a dry gas meter to monitor the coal gas flow and then a governor to manage the balance of gas line pressure to make sure adequate head pressure to maintain a steady flame at the end to each of the customers. There was a small gas meter installed for each of the customers so that a gas bill could be calculated based on their actual uses of the coal gas. Inventions and improvements seem a routine in Clegg's life back then. What is shown in Fig. 4.3 had become the prototype for a utility coal gas network built from 1815 and on. Although changes and improvements were frequently made to it the principle for the coal gas making and distribution remained essentially the same.

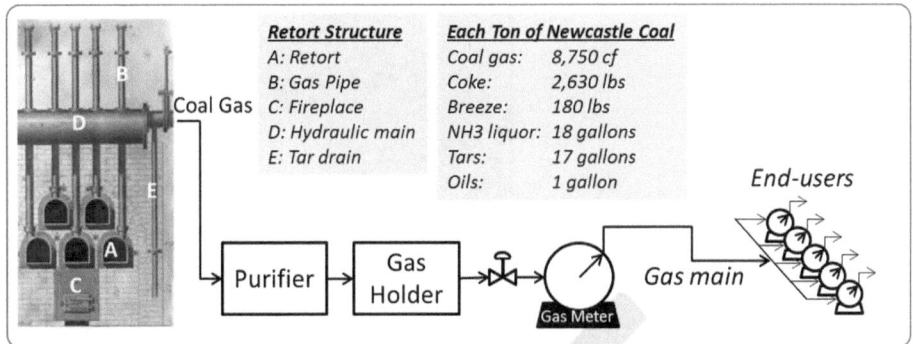

Fig. 4.3 Coal gas making system and typical products from Newcastle coal

Moving to a centralized operation for coal gas making, the design of a retort had changed from a single design (Fig. 4.2) to a bundled design that multiple retorts up to nine, called a retort bench, were housed under a heating chamber made of brick and mortar, called a retort house (Fig. 4.3). The shape of the retort changed from a cylindrical one to a D shaped one, called York D retort having a flat bottom to hold more coal while subjected to a uniform heating with the flue gas rising up from the fireplace at the bottom of the retort house. Expanding the capacity of coal gas making could be achieved by simply adding more such bundled retort houses next to each other as long as the floor space allows. When running out of space as did with the first gasworks at Peter Street, GLCC was granted to build the Beckton Gasworks in 1870 at the north bank of the River Thames at East Ham, less than 3 miles away from the Peter Street gasworks. The Beckton Gasworks processed about 3,100 tons of coal daily, manufacturing about 27 mm cubic feet of coal gas to supply to the metro London; it became the largest coal gasworks in Britain and ceased to operate until the 1960s when the North Sea natural gas came to market (Trewby). The Beckton Gasworks continued its expansion; an image made in 1881 shows its 9-retort bench house lining down the sprawling floor as far as your vision could reach, a spectacular scene during the booming time of the Industrial Revolution (Fig. 4.4). Looking at the numbers of retorts in operation and the repeating loading of coal and the discharging of coke every 8–12 h, it is not hard to imagine how labor intensive it was and how challenging the working environment was back then, which has become one of the major themes for the coal gas making down the road. Later the materials to make the retort was upgraded to clay, which increased the capacity of coal handling a bit but not necessarily change the landscape of coal gas making; the material upgrade, however, helped extend the lifetime of the retort as fireclay made retorts do not easily become distorted under high temperature and are usually more corrosion resistant.

In the early coal gas making, using the Newcastle coal as an example, one ton of coal typically manufactures 8,750 cubic feet of coal gas, which contains 51% of hydrogen

Fig. 4.4 Beckton gas works, 1881

and 24% methane, and other gases having a heating value above 500 BTU/cubic feet (Table 4.1). In the meanwhile, there are also 17 gallons of ammoniacal liquor, 18 gallons of tars and a small amount of light oils (Fig. 4.3), which had been either discharged into a river or creek or landfilled or simply piled up next to the site of the coal gasworks. For example, when Clegg put up the gasworks for the Ackerman Printing Shop sitting right next to the Thames River on the Strand into service in 1812, the liquid wastes including the spent lime water was directly discharged into the river. Soon, it raised some eyebrows from the public because of the foul smell (Samuel Clegg, 1841). In 1833, GLCC's gasworks at Brick Lane installed an ammonia plant to recover ammonia as chemicals such as muriate ammonia (ammonia chloride) for medical use; some started to spread the ammoniacal liquor directly to a pasture land (Guide G., Gas Light and Coke Co.). Some gasworks attempted to recycle the collected tars back to the retort in a hope to make more coal gas but only found out that the result was very marginal. Although some gasworks erected distillers to make use of the tars such efforts were very limited until William Perkin discovered a way to make a dyestuff from coal tars in midd-1850s. In 1851 a major improvement to the purification of coal gas took place; the Imperial Gas Co. in London introduced iron oxide to remove sulfuretted hydrogen and then used the spent oxide to make sulfuric acid with the Lead Chamber Process where pyrite or elemental sulfur were often the raw materials in the next century. The GLCC also made sulfuric acid by this process at its Beckton gasworks from 1880 until its closure. With coal gas making business rapidly expanding, gasometers had become a landmark standing big and tall in many places in Britain; some of them are still present as a showcase of the past (Thomas, 2020).

From this point on, the coal gas making process then remained largely unchanged except its scale continued to expand until 1880s when regenerative processes emerged.

In the meantime, coal gas lighting went quickly beyond the British border.

Table 4.1 Coal gases from different coal gasification processes

	I	II	III	IV	V	VI	VII	VIII
Gasification	Retort[a]	Water gas[a]			Texaco coal gasification[b]			Shell coal gasification[c]
Design	Horizontal	Vertical	Blue water gas	Carburetted BWG	Radiant	Radiant	Quench	Radiant
Coal rank	Coal	Coal	Coke	Coke	Illinois #6	Illinois #6	Illinois #6	Illinois #6
Demo project	Commercial operation				Oberhausen-Holten	Cool water	Cool water	Deutsche shell
Time	1807	1903	1875	1875	1980	1984–1989	1984–1989	1978
Coal gas composition, vol%								
CO_2	2.1	3	5.5	3.4	21.3	15.48	18.89	0.8
O_2	0.4	0.2		1.2				
CO	13.5	10.9	37.3	30	39.5	44.88	42.77	65.0
H_2	51.9	54.5	47.6	31.7	37.7	38.46	37.9	32.1
CH_4	24.3	24.2	1.2	12.2		0.16	0.04	0.0
N_2	4.4	4.4	8.4	13.1		1.02	0.4	0.7
Illuminants	3.4	2.8		8.4				
$H_2S + COS$					0.9	0.03		1.4
Total	100.0	100.0	100.0	100.0	99.4	100.0	100.0	100.0
Btu/cf (HHV)	520	532	287	540		270.5	260.5	
Specific gravity, kg/m^3	0.42	0.42	0.57	0.64	0.90	0.83	0.89	0.83

Source of data [a] Liebs (1985); [b] Corp R. (1983, 1990); [c] McCullough (1980)

Spreading of Coal Gas Lighting

Europe Before the 1820s, coal gas lighting was mostly a British phenomenon as its industrial revolution had been rolling on which made Britain the economic power far ahead of any other countries. Facing the unsettling geopolitical environment at the time, both domestic and regional, its European neighboring countries on the continent were by large still an agricultural land with limited industrial development in terms of scale. The continent, however, had started to enjoy the luxury that the new technology of coal gas lighting provided early on for street lighting but in a small foot print, only limited to a few places such as Paris, Brussels and Amsterdam and Hanover; but it grew slowly though. For example, street lighting with coal gas lighting started in Brussels in 1818, Amsterdam and Rotterdam in 1822, and Hanover in 1826. An interesting fact is that most of these developments were run and funded by British engineers and capitals. The Imperial Continental Gas Association founded in 1824 in London was designed to develop the continental market. The earliest coal gas works in Hanover was started in 1825 and put into operation the following year. In the end of 1855 German Continental Gas Company was founded in Dessau with an initiative led by Viktor von Unruh, an entrepreneur, and Louis Nulandt, a local banker, aiming to expand the local market. After lighting up streets of Dessau from 1856 onwards, more gasworks were followed and put into operation manufacturing coal gas to light up cities of Mönchengladbach, Magdeburg, Frankfurt, Mülheim, Potsdam, Lemberg, Warsaw, and Krakow on the continent.

Situation in France was a bit different remembering the pioneering *Thermolamp* invented by LeBon during the early days of coal gas lighting. Frederick Winsor who championed the GLCC in London a few years back was forced to leave GLCC around 1815 and came to Paris. He seems still ambitious to repeat what he did in London, developing a similar coal gas utility business for lighting for Paris. His efforts, however, failed in the end. The French government instead developed its own prototype for coal gas making in Paris in 1817 and provided coal gas to a hospital (*the hôpital Saint Louis*) for lighting. The government set up its own public company in 1818 to develop the market but in a regulated manner in order to prevent the unnecessary competition from the beginning (History of Manufactured Fuel Gases). By 1870, about 340 gasworks had been built to manufacture gases by using a variety of resources including coal, wood, peat and other hydrocarbons for lighting, heating and other uses (Coal Gasification). As a result coal gas had soon reached everywhere in cities and towns of fast growing.

North America To the west across the Atlantic Ocean American, after becoming independent, had enjoyed a period of rapid growth in both economy and population. According to US Census started from 1800 there were only six (6) cities or towns with population over 10,000; the number then rapidly increased to 63 cities in 1850, and 99 cities another decade later. In 1860 the largest five cities were the City of New York, Philadelphia, Baltimore, Boston, and New Orleans where the City of New York including Brooklyn had a population more than a million, followed by Philadelphia of more than a half million.

Cities, towns and villages, big and small, all enjoyed a significant inflow of population and subsequent increase of social and economic activities during the period of time, especially the following several decades post the Civil War. These places, naturally, faced the similar social challenges as England had experienced back then; the demand for a better public lighting became a reality.

In Baltimore, Rembrandt Peale (1778–1860), the American artist and museum proprietor, upon returning from his trip to Europe, immediately set up a showcase to light up his museum in 1816 with inflammable gas. In the following year, he received a charter to set up the Gas-Light Co. of Baltimore to manufacture and distribute gas for public lighting for the city. The first gasworks in America was sited at the corner of North (now Guilford Avenue) and Saratoga Street; a local technology was deployed to manufacture the inflammable gas, which was successfully used to light up the first public building, the Belvedere Theater, right next to the gasworks and two other private buildings on North Charles Street. When the first public street on the corner of Market and Lemon Streets was lighted up on Feb 7, 1817 the local newspapers recorded following (Today In Science History).

> ......that the effect produced was highly gratifying to those who has an opportunity of witnessing it, among whom were several members of the Legislature of the State from who the charter of the company was granted.

Such a gas-lighting was well received by local residents, similarly to what had happened in Britain, as a better lighting source than tallow or candle lights. By Feb. 16, additional 27 gas lights were added to the same street. It appears that what had lit up the streets of Baltimore was not coal gas but an illuminating gas, which was manufactured from distilling oils or tars (Wells, 2013). The coal gas making in US does not appear to start until around 1821 and 1822 when the Baltimore gasworks finally switched to a technology well proven in Britain using coal as feedstock. After all, Baltimore and many other cities in US have easy access to large indigenous coal reserves, which were cheaply available. Then Boston in 1822, New York in 1823, Philadelphia in 1835, Cincinnati in 1837, Washington DC in 1848, and many more set up their own coal gas making companies to develop the coal gasworks, respectively, to manufacture coal gas for the public lighting. Soon in the coming years and decades, many more cities and towns along eastern coast and the Mississippi River followed the suit.

Although America lagged behind the Britain with coal gas making for about two decades (Fig. 4.5), its deployment picked up rapidly. Based on survey of US Circa 1860, at the onset of the Civil War, every state in the Union except for Arkansas was home to at least one manufactured gas plant; the State of New York led the nation with 61 gasworks operating within its borders, followed by Pennsylvania with 48 and Massachusetts with 45. There were about 396 coal gas works installed and 431 companies involved in gas business, covering most cities with a population over 10,000 people. The New York City had four coal gas companies, each of them had established their own coal gasworks

Fig. 4.5 Chartered Gas
companies in UK and US

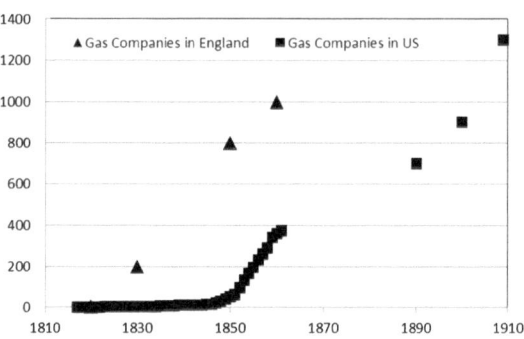

to manufacture coal gas for lighting in their respective districts (Murray, 1863) (Manufactured Gas). The states of the New York, Pennsylvania, Massachusetts, Ohio and the
New Jersey collectively held almost half of the facilities due to their extensive industrial
activities. By 1909, the number of coal gas companies went up to near 1300. Technologically, prior to 1870s most of the coal gas productions in the United States were based on
coal retort or carbonization where coal is similarly heated up in a closed and air depleted
environment; but from around 1875 the United Stated moved on with its own technology
developed in Pennsylvania. Interestingly, what was happening in America with coal gas
making pretty much mirrored that in Britain, a rapid growth following a learning process to master the coal gas making technology. A speech made by Mr. Oscar Steele, the
Superintendent of the Buffalo Gas & Light Company, at the 1873 America Gas & Light
Conference provided a good glimpse into the coal gas business and its impact on public
lives in America. Following is a paragraph taken from the speech.

> ......The chief motive for proceeding with the enterprise was public spirit, an earnest desire
> to obtain what was concerned to be a great public improvement. The science of the business,
> or the expectation of ultimate profit, had little to do with it for several years after its com-
> pletion.....But a few years changed the relations of the business. The change from the oil,
> camphene and candles was very pleasant....

Among the coal gasworks established, the Philadelphia Gas Works stood out by eventually evolving into the largest coal gasworks in America. Although the city's Second
Street between Vine and South Streets was first illuminated with coal gas on February 10,
1836, its interest in coal gas lighting actually started at least twenty years ago when its
fellow citizens, James McMurtie and Dr. Bollman proposed to the city on coal gas lighting based on their observation at gasworks in England. After a couple of reviews over the
years the City of Philadelphia finally announced an Ordinance in March 1835 to a group
of private investors to develop the coal gasworks at the 23rd and Market Streets. Samuel
Merrick, a prominent engineer and engine builder, was made the superintendent and chief
engineer to design and build the gasworks. The gasworks was under the supervision of
the twelve trustees. In designing the works Merrick adopted 10 retort benches with each

holding three retorts, a bit different from the typical 5 retort bench in England at the time, to process local bituminous coal. The gasworks includes an office, a retort house, a lime purification house, two 35,000 cubic feet gasometers or gasholders, a meter room, and a laboratory. Coal gas was first produced on February 8, 1836 and forty-six pole lamps on Second Street were lighted up successfully two days later. The success was met with more demand of coal gas so that the gasworks went through several expansions. By 1850, more equipment had been added to the gasworks; as an example, 9 more gasholders had been installed to hold its increasing coal gas holdup capacity to 1,680,000 cubic feet daily. Soon, the old site had become packed. The city then took over control of the gasworks since 1841. The chief engineer of the gasworks, John C. Cresson (1806–1876) who later became the president of the Philadelphia Gas Works, proposed to the trustees to select another site to further expand the coal gas production. The Point Breeze, about two and a half miles south of the original gas works, sitting on a lot of 75 acres next to the Schuykill River was then purchased to host the expansion. The expansion project commenced in 1851; two retort houses of an improved retort design were deployed where each house held 72 retort benches. The first retort house started to manufacture coal gas in 1855 and the next two years later (Binder) (Jackson, 1983).

In Washington DC, residents, tired of the smell and dangers of candles and oil lamps, petitioned Congress to establish a gas company to light the city. In 1848, Congress agreed to set up the Washington Gas Light Company; gasworks opened on the Mall in 1852 where the *National Museum of the American Indian* stands today. Coal gas from the gasworks illuminated lamps on the Pennsylvania Avenue and the White House, as well as local homes and businesses. During the Civil War the gasworks provided its coal gas to fill up balloons for Union Army's reconnaissance Corp. The gasworks remained there for over 50 years.

Asia Pacific Coal gas lighting did not arrived in Asia Pacific until 1856, pretty much following the steps of the Western expansion into the regions, and reached some countries such as New Zealand, India, Hong Kong, China, Singapore, Sri Lanka, Japan, Thailand, Jakarta over more than half a century (Fig. 4.6). Scales in these countries are typically small due to limited industrial activities. The coal gas works in Shanghai were one of the earliest few developed in the 1860s and later became the largest in scale in the Far East region, producing 700,000 cubic feet (19,800 cubic meters) of illuminating coal gas daily by 1900 after a series of upgrading and expansions.

According to the Office of Shanghai Municipal Chronicles, the early work to build coal gas works was initiated by the British Concession who posted a tender in Daily Shipping List Commercial News on February 26, 1860, aiming to raise 100,000 Liang of silver during the late Qing Dynasty. Silver had been one of the popular currencies and one Liang (俩) is 50 g today in China but was about 31.25 g prior to 1949. Shanghai in the next few years witnessed the siting and constructions of gasworks, gas mains, and street lamps. A land of 9,876 m^2 on the south bank of the Suzhou River, current Xizang Zhong Lu was sited for the coal gasworks. Gas mains and pipes (51–229 mm in

Fig. 4.6 Years of coal gas
works established in countries
of the Asia Pacific (data from
Hatheway website)

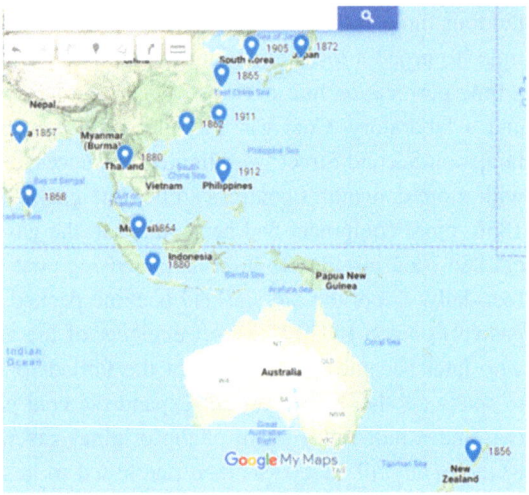

diameter) totaling 5,051 m were laid from the coal gasworks to the Bund at the Nanjing
Road East via the Zhejiang Zhong Lu. Coal gas retort was a typical design at the time,
one five-retort bench followed with gas purification, a gasometer of 1,700 cubic meters
and coal gas distribution system. The coal gasworks, the British Concession Gas Works,
was completed in September 1865 and began to dispatch coal gas two months later. The
coal gas replaced about 205 kerosene lamps on the Nanjing Road, the popular commercial
street in Shanghai, plus 185 private buildings along the street. Coal gas was sold at 4.5
YinYuan (Silver Dollar) per thousand cubic feet and the rental cost for a gas lamp was 5
Silver Dollar each per month. Following the British Concession, the French Concession
in 1867 built the second coal gas works in its territory, current YanAn Dong Lu between
Yongshou Lu and Guangxi Nan Lu, producing coal gas to phase out the kerosene lamps
on its streets. Shanghai kicked off its rapid expansion and growth of the coal gas business.

In 1878, the British Concession Gas Works expanded its operation by adding five
more horizontal benches while upgrading its purification by replacing the previous slick
lime process with the iron oxide process. The expansion increased the capacity of the
illuminating coal gas and the distance of gas mains and pipes by 6 times. Of significance
at this time is that coal gas also became used as fuel gas for domestic cooking, creating
additional demand for coal gas. To meet such a demand the British Concession Gas
Works acquired the French Concession coal gasworks in 1888, and also added four more
horizontal benches to its original gasworks in 1893. By now there were total of 11 retort
benches in operation manufacturing about 13,000 cubic meters of coal gas daily. In 1900
the British Concession Gasworks was reorganized in Hong Kong as the Shanghai Gas
Co. Ltd. and the Kadoorie family became its major stakeholder. In the same year, the
Shanghai Gas Co. upgraded two of the 11 horizontal retort benches with three sets of
a regenerative horizontal retort system. The more efficient regenerative horizontal retort

system had increased coal gas rate to 13,900 cubic meters daily, making the Shanghai Gas Works the largest coal gas manufacturer in the Far East. Between 1900 and 1934 Shanghai Gas Co. revamped its gasworks by exploring new technologies including vertical retort ovens and water gas generators etc. After relocating its coal gas operation to a new site in 1934, the company's coal gas capacity had augmented to 113,300 cubic meters of coal gas daily and serving 13,400 customers via its 190 miles of gas mains. When WWII broke out in the Pacific on December 8, 1941, however, the Shanghai Gas Co. lost its control of the gasworks to the Japanese Army and did not get it back until the end of WWII in Asia Pacific (City, 2023; 张应莹, 1960). The city of Shanghai serves as one of best epitomes when looking into the evolution of coal gas making in the Fast East.

References

Accum, F. (1815). *A practical treatise for gas-light.*

Accum, F. (1819). *Description of the process for manufacturing coal gas.*

City, S. (2023). *Shanghai coal gas.* Retrieved 2023, from Shanghai City Chronicles: http://www.shtong.gov.cn/Newsite/node2/node2245/node4516/node55027/index.html

Corp, R. (1983). *Environmental characterization of the texaco coal gasification process at the Ruhrkohle/Ruhrchemie in Oberhausen-Holten, FDG.* EPRI.

Corp, R. (1990). *Cool water coal gasification program: Final report.* EPRI.

Hughes, S. (1853). *A treatise on manufacturing and distributing coal gas.*

Jackson, D. C. (1983). *Historic American engineering record, Philadelphia gas works.* National Park Service, US Department of Interior.

Lawrence Liebs. (1985). Town gas—An overview. In *AGA distribution/transmission conference* (pp. 1–43). AGA Distribution/Transmission Conference.

Long, R. (2022). *May 24, 1935 under the lights.* Retrieved Jan 31, 2023, from Today In History: https://todayinhistory.blog/2022/05/#:~:text=The%20Westminster%20Review%20newspaper%20opined,Baltimore%20lit%20up%2C%20in%201816

Major, J. (2018). *All things Georgian.* Retrieved Mar 2023, from The Grand Jubilee of 1814: https://georgianera.wordpress.com/2018/07/31/the-grand-jubilee-of-1814/

McCullough, G. R. M. v. (1980). Shell coal gasification process. In *Fischer-Tropsch archives* (pp. 41–64).

Murdock, W. (1808). *An account of the application of the gas from coal to economic purpose.* Royal Society.

Murray, J. B. (1863). American gas works. *The American Gas-Light Journal*, 370–373.

Rowlandson, T. (1809). *A peep at the gas lights in Pall Mall.* Retrieved June 2023, from Science History Institute: https://digital.sciencehistory.org/works/6w924c79b

Samuel Clegg, J. (1841). *A practical treatise on manufacture and distribution of coal-gas.*

Thomas, R. (2020). *The manufactured gas industry: Vol. 3 Gazetteer.* Historic England.

Wells, B. W. (2013). *Con Edison.* Retrieved March 17, 2022, from American Oil & Gas Historical Society: http://aoghs.org/stocks/con-edison-american-utility-company

张应莹. (1960). 煤气机与煤气炉问答. 上海: 上海科技出版社.

Chemistry and Coal Gas Making

5

Having overcome the learning period coal gas making proved itself a safer and cheaper alternative as a public utility for lighting compared to the traditional candles and oil lamps. By 1850, there were about 800 coal gas companies established in Britain among which 13 were in London. Although there was no rival technology in the field of coal gas making the increasing number of players rushing into the field had created an intense competition that had eroded the profit margin for owners and stakeholders. In many cases, the competition had compromised the quality of coal gas and related services. To protect consumers from malpractice by business owners, in the meantime, Parliament stepped up its efforts to regulate the market for public lighting. The Gas Regulation Act in 1820 started to regulate the sales price of coal gas tied to the heating value of the coal gas sold, and the quality of the coal gas ought to be monitored; an example is that the coal gas at the gas main shall not be less than 2″ of water pressure and so on. More Parliamentary acts were issued after 1840 to maintain market order as well (Meade, 1921). The fierce competition and the increasing regulations had forced the coal gas making operators to look for ways to drag down the cost of their coal gases while continuing to be able to afford of paying dividend to its stakeholders in order to attract more capital.

In the meantime, chemists, inventors, and entrepreneurs were motivated to look for ways to either improve current coal gas making processes or other options to upgrade the existing technology for coal gas making. Such efforts did lead to the creation of some new schemes and processes developed.

© The Author(s), under exclusive license to Springer Nature Switzerland AG 2024   43
Q. Zhuang, *From Coal to Hydrogen*, Synthesis Lectures on Chemical Engineering and Biochemical Engineering, https://doi.org/10.1007/978-3-031-55586-2_5

Status-Quo of Chemistry

What Accum and Clegg contributed to the early success of GLCC's business represents two different scientific disciplines, applied chemistry and chemical engineering. Back then a successful execution of gas works pretty much depends on those individuals who were hands-on by closely involved in the development of coal gas works. This holds true even more than 200 years later today that any coal gasification projects are still by large a state of the art; hands-on experience is a must.

Entering into the nineteenth century, Lavoisier's oxidation theory had gained more public acceptance in general except for the fact that it took a few more decades to clarify some glitches such as that the caloric is not an element, and how it works with heat in the environment to make coal gas. The fact that the oxidation theory was able to provide a simple, consistent, and convincing explanation about the combustion of coal, coal gas, candle, and any other inflammable substances somewhat facilitated the deployment of coal gas for lighting business, at least, from the viewpoint of demystifying about fire and its associated fears. The impact of chemistry to coal gas for lighting at the time would appear more philosophical than practical or applied. To advance the modern chemistry, Humphrey Davy played a significant role in advancing and promoting the chemistry. Although taking a competitor's approach against Lavoisier's theory Davy's work and discoveries, interestingly enough, not only left Lavoisier's chemistry system largely intact, but to the contrary confirmed its soundness. Specific to caloric, however, Davy demonstrated that caloric was neither a substance as an element nor tangible nature. By applying battery technology pioneered by Alessandro Volta (1745–1827), Italian physicist, Davy was able to extended Lavoisier's table of elements by adding at least nine more elements. Also through giving public lectures, his exciting, sometimes, bizarre, and risky demonstrations of chemical experiments certainly entertained the public interest and curiosity toward science, chemistry, and their roles in improving public lives at that time in Britain.

With more elements discovered questions of what matters are made of, what coal is, and how the coal changes during coal gas making and so forth had naturally become the subject of many investigations and studies in the early eighteenth century and continued throughout most of the century. When Murdock. In his presentation of the first coal gas work at the Phillip & Lee's mill to the Royal Society in 1808, Murdock was candid that he was not familiar with what was going on in the fields of chemistry and pneumatic chemistry during his time of experiments in the 1790s; nor was he familiar with all those gases that evolved from retorting coals except for the potential as an inflammable gas for lighting. That is probably how things were evolving back then when chemistry was little know, and when knowledge and skills were learned and accumulated by practicing. From this angle, the knowledge that Murdock had learned about coals that were used to make coal gas might well not be much different from what Hales knew back in the 1720s. Here is what Accum stated in his 1819 treatise about what coal and coal gas are.

All substances, whether animal, vegetable, or mineral, consisting of carbon, hydrogen, and oxigen, when exposed to a red heat, produce various inflammable elastic fluids, capable of furnishing artificial light.

The gases thus obtained are called carburetted hydrogen they produce, from their combustion, water and carbonic acid. The species of carburetted hydrogen, procured from pit-coal, has of late been called coal gas.

The pit-coal is a bituminous coal, mined from Newcastle and Whitehaven areas in Britain at the time, which gives off the most coal gas in coal gas making. In 1841, Clegg Jr. in his treatise cited the following analysis in percentage of the Newcastle pit-coal.

Carbon	75.28
Hydrogen	4.18
Nitrogen	15.96
Oxygen	4.58
Total	*100.0*

This becomes probably the earliest elemental analysis of coals recorded for coal gas making. There is no need to dig into those numbers as the understanding about coals at the time must have been limited simply by looking at the nitrogen number only, unrealistically high.

Dalton's atomic theory proposed in 1803 seems to have attracted many including those involved in coal gas making. Dalton was also consulted as one of the go-to persons for Clegg when facing fundamental problems and questions regarding coal gas and its by-products.

The Perspective of Engineering

Coal gas making by retort is an external heating process, getting the heat from coal firing through the wall of retort to heat up coal inside. Heat transfer in the retort setup, however, is a very slow process, which is why a batch or cycle operation takes 8 to 12 h to accomplish. Furthermore, it is an energy consuming process as well in that firing up the retort to desired temperature takes as much as 50% of the coal charge to retort; the heat loss in the cycle operation is also paramount with the heat remaining in flue gas lost to chimneys and in the discharged red hot coke lost to the surrounding environment by water quenching. In short, it is a lousy operation of low efficiency and low yield, which resulted in a high consumption of coals. Coal was not cheap at the time. To change that, there had been many engineering attempts made to get more coal gas out of the existing retort operation by leveraging the red hot environment in the retort or the red hot coke discharged from it at end of each cycle of operation. The chemistry of coal gas making then started to shift away from the distillation or pyrolysis principle to chemical reactions by applying the old chemistry discovered by Fontana in 1780, water reaction with hot coke, or air oxidation to give off inflammable gas.

One of the early state of arts was a device filed in 1824 by British inventor, John Ibbertson, who suggested injecting steam into a bed of incandescent coke or charcoal, heated externally with a coal fireplace, to produce illuminating gas, and the steam would react with carbon in the coke or charcoal to give off hydrogen, carbonic oxide, and carbonic acid. In 1839, another British inventor, Alexander Cruckshanks, proposed to inject a superheated steam into a vertical retort, sitting in the middle of a coal fireplace, filled with carbonaceous materials. The superheated steam was generated with flue gas of the fireplace as indicated in Fig. 5.1L. The inventor in his patent claims *"(1) The heated gas from the retort and the hot air from the furnace are applied to generate and to heat the steam, as also to heat the water with which the boiler is supplied. (2) that the steam is introduced to the retort at the temp at which it is decomposed by carbonaceous substances, whereby retorts of much large dimensions may be used than with steam at lower temp."*

To utilize the residual heat in the coke discharged from retort at the end of each cycle operation, Mr. George Lowe in 1831, the chief engineer of the GLCC, transferred the red hot coke to a specifically designed iron cylindrical column covered with a lid. The column was brick-lined inside for insulation to protect the iron column. During the transfer the coke would burn quickly and get hotter upon exposure to the open air; after the transfer, the lid closed immediately and followed by injecting steam from the top of the column and the produced gas exited from the bottom. Such an intention was to utilize the residual heat in the red hot coke to generate some additional coal gas or better called water gas. These attempts share one common feature, that is, that the heat required necessitating the water-carbon reaction is not sustainable and has to be provided externally with coal firing because the water-carbon reaction is a strongly endothermic reaction, requiring a lot of

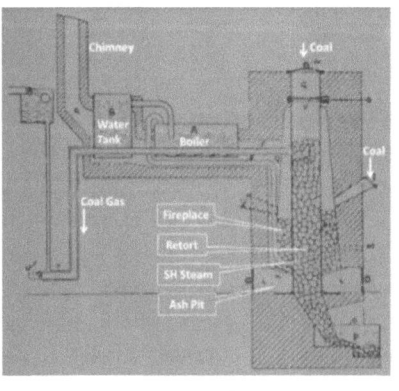
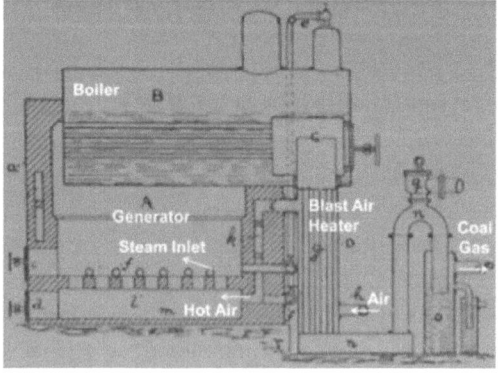

Left   Right

Fig. 5.1 Cruckshanks' vertical retort water gas apparatus (BP Patent 8141, 1839) (L), and kirkham brother proposed apparatus (Special Report on Water Gas, 1886) (R)

heat to maintain its temperature. What Lowe did to salvage the residual heat that remained with coke seems to make sense from an energy utilization perspective, but would hardly be justified by the additional investment of the equipment and the hazardous operation environment for the small amount of gas made, which tells why these attempts of the state of the art in nature did not get a go on an industrial scale. So true to many other similar trials attempted as well at the time such as co-injecting steam, either saturate or superheated, with oils such as rosin and tallow oils over a red hot body (coke or wire etc.) in a similar setup to generate illuminating gases (British Patent No. 11654 by Stephen White, 1847; US patent 21,027 by J. Milton Sanders, 1858) (Special Report on Water Gas, 1886).

To overcome the difficulty in obtaining heat for the process, two brothers, John and Thos Kirkham, proposed in their 1852 patents (B.P. No. 14,238, 1852) to inject preheated blast air and steam through a coal bed, which is heated to almost a melting state of iron to produce an inflammable gas containing hydrogen, carbonic oxide, carbonic acid and a large amount of nitrogen (Fig. 5.1R). In this way, air oxidation would provide the required heat internally for the water-carbon reaction to proceed. The process also uses the waste heat in the product gas to heat up water for steam generation and to preheat the blast air. As one of the earliest state of the arts, such a delicate embodiment not only uses an efficient internal heating but also recovers waste heat for reuse to generate steam. One inevitable of doing so, however, is a large amount of nitrogen coming in with air compromising the quality of the illuminating gas for lighting. Even though they improved the design in 1854 (B.P. No. 1882) by adding a heat chamber to have the product gas carburetted, the effect seems limited in obtaining both the wanted heating power and illuminating power.

Well, the efforts in getting more inflammable gases in a retort setup seem to hit a wall. Considering the inherent features of the retort system, the external heating, a less efficient and time consuming process, seems to blame. It is the chemistry taking place inside the retort that in fact dictates the approach, engineering, of the gas making. It appears that coal gas making through retort had reached an end of no-where to go. Now, would a change in the engineering warrant a better chance for the new chemistry? What was happening in neighboring fields came to a rescue.

By 1800, the blast furnace technology in Britain had advanced significantly, churning out 130,000 tons of iron products annually, thanks to the coke that Darby introduced in 1709 for iron making and the more powerful bellow that made the large scale operation possible. In 1861, Britain produced 3.8 million tons of iron. Iron making with blast furnaces had become well established. Ironmasters must have acquired the required techniques and skills as well to make iron by layering iron ore, charcoal or coke, and limestone, one on top of the other, inside the blast furnace and then firing it up to as high a temperature as possible. Hours later, molten iron would seep out at the bottom of the furnace. In the meantime, there was a significant amount of gases, inflammable off-gas, left at the top of the blast furnaces. For instance in 1837, Wilhelm von Faber du Faur successfully used the off-gas to make a wrought iron from a pig iron in a reverberated furnace at Wasseralfingen in Wuertemberg. According to a report made to the French Academy of Sciences on May 28, 1842, Frenchman M. Audertot first started to use the inflammable off-gas discharged from a blast furnace to roast iron ore and fire limestone around 1809 and 1811; he also took a patent for it (Wyer, 1906) (Albrecht Kaupp, 1984) (Roehrig, 1870). The fact that the blast furnace off-gas is inflammable must have been known well before that time as this was the time on the other side of the Atlantic when Murdoch and Clegg had just started their early commercial scale coal gas works and were trying to put more into operations to produce the inflammable gas for lighting. Between Faber and Aubertot, there were more trials that had been made and quite a number of patents had been issued about the utilization of the off-gas for metallurgy applications. With regard to what's happening inside the blast furnace or the chemistry on the formation of the off-gas, however, little seems known although there appears a certain level of understanding about the complicated nature of the chemistry. This can be seen from a paragraph from a book published in 1870, "*A Practical Treatise on Metallurgy*" by Crookes and Roehrig.

> In the employment of the waste gases (off-gas), it was, however, soon discovered that a modification in the process of the blast furnace modified the waste gases with regard to quality and quantity, in consequence of which the fining process became impaired. The collection of the gases also seemed to injure the process of the blast furnace.

The fears over the potential impact on the blast furnace operation might have prevented further exploitation of the inflammable blast furnace off-gas. After all, the objective of a blast furnace was to produce the most wanted iron products but not the off-gas. The realization of the off-gas or fuel gas as a better way for industrial process heating, however, had created a strong interest in developing options outside the blast furnace to produce fuel gases. There had to be something new and here comes a new chemistry to play.

Around 1839–1840, under the background that there existed a broad interest in finding the use of the waste gas from blast furnaces around iron making, Gustav Bischof, a German chemist, and M. Ebelmen, a French chemist, worked independently to invent apparatus for generating fuel gas by reacting carbon with air (oxygen in the air) or air and steam. They called it gazogene in French or gas producer in English, a stand-alone apparatus independent of the blast furnace operation. While there seem conflicting accounts on who first invented the gas producer and applied it to the iron works of Audincourt of France there were some extensive trials at the works carried out around 1840 on an industrial scale. It seems that several types of gas producers were deployed to generate fuel gas for steel making (Wyer, 1906) (Albrecht Kaupp, 1984). Figure 5.2 shows two examples, a slagging gas producer invented by Ebelmen, similar to a blast furnace design and operation, and a dry ash discharge producer by Bischof. The slagging producer uses charcoal or coke fed from the top through a cast iron column that reacts with blast air coming in from the bottom to produce fuel gas, which consists of primarily carbonic oxide and some carbonic acid. The resulting gas is discharged from the top left corner. The cast iron column is always filled with charcoal as a seal preventing leak-out of fuel gas or a lid could be used instead on the top. To assist ash discharged in a molten state at the bottom of the producer a small percentage of iron slag and clay are co-fed as fluxant with charcoal or coke. In the other producer, the dry ash type gas producer is placed right next to a puddling furnace to provide fuel gas to it. The gas producer is a rectangular brick–mortar design with solids fuel charged from the top and sitting on a grate where blast air blows in from beneath the grate. Steam is injected into the lower part of the coal bed from an opening right above the grate. It is interesting to notice that the fuel gas left the producer at the upper right corner flows directly to a puddling furnace and there joined by a stream of air preheated by utilizing the waste heat of the puddling exhaust. Obviously, this type of gas making had adopted a new way to obtain heat to heat itself up; it is also heated internally from the reaction of carbon with air (oxygen in the air), a much fast and more effective. The inflammable off-gas from the system is quite different from coal gas in many ways such as heating value, and compositions etc., not suitable for illuminating purposes. Furthermore, what happens inside the gazogene is also completely different as well and features a new chemistry, to completely convert coal or coke into inflammable gas, which had not yet been understood chemically at the time.

At the time gas producers were in general built with brick–mortar structure, similar to the puddling furnace, countercurrent flow (with gas media going up and solids moving down) and in continuous operation to manufacture fuel gas. The puddling furnace (Fig. 5.2 Right) with a built-in dry ash gas producer also shows an example of using the puddling exhaust flue to heat up cold air for fuel gas combustion in the heating chamber over the pool of iron (E area).

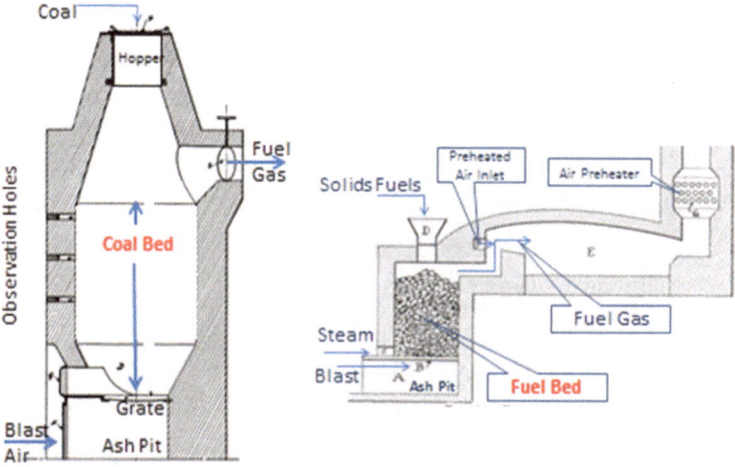

Fig. 5.2 Gas producers of slagging type (Left) and Dry-Ash type (Right) (Wyer, 1906)

Despite the difference in ash discharging mode, slagging vs. dry, the two designs essentially share the same principles of chemistry and engineering. Using coal as an example, what happens within the producers is a combination of physical and chemical processes, combustion, reduction, pyrolysis and drying, taking place in the sequence from the grate or bottom to the top of the coal bed in a producer (Table 5.1). During actual operation, blast air moves up along with steam or without, countercurrent to coal moving down. In moving up air (oxygen) reacts with carbon to produce carbon dioxide and releases heat to heat up to red hot, above 1200 °C, in the combustion zone; carbon dioxide along with nitrogen moves into the incandescent zone where carbon dioxide gets reduced to carbon monoxide while losing part of its heat; then the resultant gas carrying the remaining heat goes into the pyrolysis or distillation zone to cook out the volatile matter in coal while losing additional heat; finally, the residual heat in the final fuel gas heats up raw coal fed from top to drive away its moisture. Emerging from the coal bed into an open space, the fuel gas along with moisture, and the dust of fine particles entrained within coal bed leaves the producer at about 150–200 °C toward the regenerative system. Obviously, the heat exchange within a producer is very effective compared with a retort operation. On the reversing move, coal once being fed into the top of the producer would subject to heating and start to lose its moisture; the dried coal moving down further begins to lose its volatile matter and gives coke; the resultant coke reacts with carbon dioxide in the incandescent zone to form carbon monoxide and becomes consumed; the residual carbon then gets burnt up in the combustion zone to release heat, both sensible and radiative, to achieve a complete consumption of coal. Ash, molten or dry gets discharged at the bottom of the producer. Relative to the ongoing coal gas making, the invention of the gazogene or the producer does provide drastically different characteristics in terms of effective use of heat and complete destruction or conversion of coal feedstock in producing inflammable gas fuels that would revolutionize the coal gas making, which had so far been using the conventional retort operation. Such a revolution, however, did not happen at the time except for some limited uses in metallurgical processes, which operate at a less extreme temperature. Nitrogen this time in the inflammable fuel gas seems to blame.

As indicated in Table 4.1 of Chap. 4, fuel gas generated from producer setup contains more than 50% of nitrogen due to the use of air as blast. The resulting fuel gas is of a low heating value ranging from 110 to 150 BTU/cubic feet. Such a fuel gas, however,

Table 5.1 Zone distinction within a producer

Zone	Temp, °C	Primary reactions	Chemistry
Drying	Up to 350	Driving out moisture	Physical process
Distillation	350–900	Similar to retort	Pyrolysis
Incandescent	900–1200	CO, H_2, CH_4	Gasification
Firing	> 1200	CO_2	Combustion

would not be able to help metallurgists achieve the temperature highly sought after at the time in making glass or steel of good quality. Well, what about using oxygen in place of air in a producer operation? It would help for sure except the fact that the invention of oxygen separation technology was not available until a half century later.

Compared with the current coal gas making process and those trials relying upon external heating, technically speaking, the advantages of a gas producer operated by injecting air, steam, or a mix of them into the solids hydrocarbons bed seems obvious. The producer design serves as an effective apparatus that accommodates such a complicated physical chemical process to utilize coal resources for the production of fuel gas. It provides a nonstop flow of fuel gas in a continuous operation through a much more effective internal heating. The potential value of such an invention appears paramount and no other processes so far could match. It is definitely a leap in engineering assisted by the increased knowledge of chemistry. At this point, it appears that finding ways to unlock the potential value of such a producer had become the focus of scientific and industrial activities. What solution could provide a breakthrough to unlock such a value? The approach seems logically to point to the nature of a temperature and factors behind it, the science, understanding the fact that common knowledge at the time normally links heat to a temperature. A convenient analogy is that temperature of water rises when water subjects to coal burning. The reality, however, is that little was known back in 1840s about what is heat and how the heat works to raise a temperature. Plus, if any, Lavoisier's theory of caloric had been the main street of an authority even though eventually met with increasing confusions and then questions. It seems that the understanding of chemistry somehow held back the advancing of coal gas making; or in a more accurate way, coal gas making would have a hard time to achieve another milestone without understanding the true nature of caloric. This may be better examplified by what Clegg Jr. described in his 1841 treatise on coal gas making.

> The effects of caloric should be well understood by the chemist: in his hands it is an important power, by which he not only brings bodies into a proper state for combining, as when he converts a solid into a liquid by heat, but also is enabled to destroy previously existing combinations; for caloric not only counteracts the effects of cohesion, but also those of affinity, when the heating process is carried far enough. Thus we form quicklime by heating the natural compound carbonate of lime, or limestone, to redness; the affinity which unites the carbonic acid to the lime becomes neutralized or destroyed by the caloric, and in consequence the gaseous acid escapes, leaving the pure lime in the kiln.

Concurrently, an English brewer who had become obsessed with the concept of heat since young had been working enthusiastically on the subject of heat and its equivalence with other forms of energies; he in 1845 finally landed on a hard piece of evidence that others had no reason to reject about the nature of heat and its relations with temperature and other forms of energy, which provided a time window of opportunity for the gas producer to come around.

It is James Joule.

References

Albrecht Kaupp, J. G. (1984). *Small Scale Gas Producer-Engine Systems*. Springer Fachmedien
    Wiesbaden.

Meade, A. (1921). *Modern Gasworks Practice*. Benn Brothers Ltd.

Roehrig, W. C. (1870). *A Practical Treatise on Metallurgy*. Longmans, Green and Co.

Wyer, S. S. (1906). *Producer Gas and Gas Producers*. The Engineering and Mining Journal.

The Power of Gas Producer

6

With coal gas making capacity going larger and larger, the multi-retort bench module design and the housing brick–mortar structure resulted in a significant saving of fuel coal used to heat up the retort bench as long as the operation goes around the clock, which would have less heat lost to an open environment. The effort to change the chemistry in making more gas by utilizing the hot coke at the end of each retort operation had yet proven to work, both technologically and economically. The retort operation still discharges a large amount of coke which is still the source of heat loss. There seems a need to find a way to better manage the heat. Before the 1840s, the reality of what heat is and the nature of it, however, was still in a chaotic moment. What had achieved, nonetheless, is that there would be a significant saving of fuel coal if coal loading takes place while retort is still hot, which was enhanced with an integrated multi bench-retort complex built with a brick–mortar structure instead of an iron material as clay retains heat much better than iron.

Like the coal gas making industry, the railway industry had grown rapidly since the first commercial railway started running between Liverpool and Manchester in 1829. Such rapid growth had created a strong demand for better materials, stronger and more durable, to make steam engines and rail tracks. Before that time, the rail track made from wrought iron did not last long. To make the steam engine and rail tracks stronger and more efficient, steel material instead of wrought iron or cast iron had been known as a better material to serve the needs in terms of its hardness and tensile strength. To make steel, however, a high enough furnace temperature must be provided in the steel making process. The problem is that the technology to raise the temperature high enough so as to melt an iron completely simply did not exist back in the 1850s, nor did the fundamental knowledge such as on what contributes or impacts or controls the temperature. Back then, steel was

Q. Zhuang, *From Coal to Hydrogen*, Synthesis Lectures on Chemical Engineering and Biochemical Engineering, https://doi.org/10.1007/978-3-031-55586-2_6

a material of high price and typically produced in small quantity. The calls for a steel material and the need to understand the nature of heat provided a perfect opportunity for the changing chemistry in gas making to find its value and use; thereof a technology called gas producer of complete destruction of coal to produce an inflammable gas was born. It is the Siemens brothers who made it happen.

Heat, Caloric, and Energy

While creating a fire by wood friction or stone sparks had been practiced for thousands of years what was behind the phenomenon of the friction or the sparks creating the fire had puzzled philosophers and scientists; no one made the connection between the mechanical motion, the generated heat and the caused fire. In his combustion theory, Lavoisier believed the element of caloric residing with oxygen gives heat and light during combustion. Later in 1799, Davy rejected this idea with his experiment by rubbing two pieces of ice cubes to cause the ice melt at the rubbing surface. Although Davy did not provide any alternative explanation to Lavoisier's theory there seems a common belief at that time that ice only melts upon a temperature increase. A similar holds true to the steel making process, a temperature of white hot instead of cherry red hot was necessary to melt an iron more thoroughly in order to make better steel with the puddling operation. On the surface, the ice melting and the iron melting appear totally different things; in essence, they might well share the same principle.

Like what was practiced by ironmasters to recover heat from exhaust flue to preheat blast air to help raise the temperature of forging or puddling for the purpose of producing a better iron product, steel, coal gas making business was actively looking for alternatives to improve its efficiency to save coal consumption when coal prices became high. Heat somehow eventually evolved as a subject of interest that attracted physicians, physicists, philosophers and so on to investigate and study the difference between heat and caloric, which led to the establishment of thermodynamics and as well physical chemistry.

The French scientist Sadi Carnot (1796–1832) was born into a family of the French Revolutionary figure, Lazare Carnot. Having gone through a turbulent life due to the rise and fall of the Napoleon time, he developed a strong interest in sciences and their applications. During 1812 and 1814, Sadi Carnot received a fine education at the École Polytechnique ("Polytechnic School"), a public school established to advance the science and engineering of France in 1794, where military officers of the French military were trained with a focus on applying sciences such as mathematics, chemistry and physics etc. to industrial and economic problems. It is during this time that Carnot exposed himself to some great minds. His studies stress the skills of analysis, mechanics, descriptive geometry, and chemistry, which were taught by several distinguished faculties including Poisson, Gay-Lussac, Ampère, and Arago. After graduation, Carnot served as an army officer. Driven by his scientific interests, Carnot had propelled himself into the public discussions on chemistry, physics, and their applications such as with steam engines. After 1815 Britain was far ahead of France in terms of technology, application, and manufacturing of steam engines. Designing and manufacturing the steam engines required a lot of hands on expertise and information in order to achieve the desired performance and efficiency of the heat engine, either of different engine designs (single cylinder or multi-cylinder) or under different steam conditions (low pressure vs. high pressure or saturate steam vs. superheated steam). Upon closely examining the steam engines operated by the French Army and analyzing the associated works Carnot believed that the ineffective use of steam was to blame for its low efficiency and pointed out that the nature of the steam cycle in his 1824 treatise titled "***Réflexions sur la puissance motrice du feu et sur les machines propres à développer cette puissance (Reflections on the Motive Power of Fire)***". In the essay, he predicted that the efficiency of a heat engine depends only on the conditions of the working medium at extreme points during the cycle operation where heat or caloric as the medium could not be created or destroyed. The work done during expansion and condensation had nothing to do with the types of the medium, being water or other fluids. What's unfortunate is that Carnot's publication had been largely ignored until ten years later when another French engineer Émile Clapeyron helped extend Carnot's results, which became eventually accepted and formed the basis of the thermodynamics principle of modern days (Mendoza, 2023). Although Carnot borrowed Lavoisier's caloric theory to explain his theory, which had by the time proven wrong, there was neither any consensus on what heat or caloric should be. What Carnot got it right is the principle that heat or caloric can neither be created nor destroyed. Carnot's work was finally recognized by the founders of thermodynamics, the German physicist Rudolph Clausius and the British Chemist William Thomson (Lord Kelvin, 1824-1907) in the 1850s and incorporated it into the first law of thermodynamics, the law of energy conservation. William Rankine (1820-1872), a Scottish engineer, later in 1859 took a step further from the Carnot Cycle for ideal conditions to establish the so called Rankine Cycle for practical conditions to reflect the efficiency of a heat engine under real working conditions, which is that the efficiency at a practical condition is always lower than that at an ideal condition.

Similarly, the fact that the word of *energy* was first used by the British physician and physicist Thomas Young (1773–1829) in as early as 1800 went unnoticed until 1850s (Morse, 2018). In one of his lectures, Young used the term of energy to substitute the "living force" of a moving body, a product of the body mass (m) and the square of its moving speed (v), (mv^2), which was a subject of physics back then. To establish the link among heat, caloric, electricity and energy etc., however, the observations and experiments made by Mayer and Joule in the 1840s were important to connect the dots while helping understand the nature of them, paving the way towards establishing the doctrine of thermodynamics. During his trip to Java in 1842, Julius Robert von Mayer (1814–1878), a Prussian physician and physicist, noticed the redder venous blood of people living in tropical weather than those in a cold region. He made speculations that food intake, body heat, and work were somehow interconnected and based on which he developed the concept of energy conservation that body heat and body work are interchangeable, an equivalence that ought to exist ubiquitous. Mayer published his results in the journal *"Annalen der Chemie und Pharmazie (Annals of Chemistry and Pharmacy)"* owned by Justus von Liebig but also went unnoticed (Julius R. Mayer).

Different from Mayer, James Prescott Joules (1818–1889), a brewer and physicist, was born into a brewer family, received an education from John Dalton in his early age and also influenced by many great engineers in Manchester, an industrial center at the time. At a young age Joule became fascinated with electricity. In evaluating the feasibility to replace a coal fired steam engine used in his family brewery with a newly invented newfangled electric motor powered with a Volta battery using zinc as cathode, Joule uniquely presented his conclusion by estimating that one pound of coal could produce steam to do five times as much work as that with a pound of zinc battery powered electric motor and the cost benefit would be 20 times (James Prescott Joule). In the comparison, Joule adopted the concept of equivalence or a common denominator, the equivalent work done by steam engine or the battery. Battery powered electric motor was much more expensive to operate, at least. It is not hard to imagine that coal must be cheaper than zinc, a consumable in running a Volta battery. The work might obviously have propelled Joule to go further to contemplate if heat, magnetic, and electric power were somehow connected or interchangeable. Being familiar with Carnot's and Clapeyron's work in France Joule went on to set up his laboratory equipment to investigate the convertibility of various forms of energy into one another.

In 1840 Joule presented his findings at a conference with following conclusion based on his experiment with electromagnetic engine and Voltaic battery.

The mechanical power exerted in turning a magneto-electric machine is converted into the heat evolved by the passage of the currents of induction through its coils; and, on the other hand, that the motive power of the electro-magnetic engine is obtained at the expense of the heat due to the chemical reactions of the battery by which it is worked.

The findings later form the basis of what is called the Joule's Law, which also clearly indicates the equivalence of the heat generated with electricity and by the chemical reaction within the battery. In the following years, Joule moved on to demonstrate that the electromagnetic engine also allowed him to convert mechanical work to heat. By now, Joule had become confident about his assumptions that different power sources were interconnected and interchangeable. In his next experiment designed to quantify such equivalence, he started to make his name known. The presentation titled *"On the mechanical equivalent of heat"* that Joule made to the British Association in Cambridge in 1845 did create some ripple effects among the great minds at the time. In his experiment (Fig. 6.1) Joule measured the heat in a British thermal unit ("BTU") required to raise the temperature of a pound of water by 1° was equivalent to a falling weight of a pound to spin a paddle-wheel to stir the water by traveling a distance of 819 feet. To describe it in another way

Fig. 6.1 Joules heat apparatus
1845 (James Prescott Joule)
(Wikimedia Commons, 2006)

or in today's terms, the potential energy of a weight of one pound traveling 819 feet of distance is equivalent to the energy required to raise 1° of one pound of water. With more experiments, Joule was able to refine the traveling distance of one pound of matter to 772.692 feet in 1850.

In hindsight, both Meyer and Joule were able to make the connection between different energy forms and their interchangeability. Although Meyer's proposal seems a bit abstractive Joule's experiments were quite quantitative and convincing. A natural question could be how Mayer and Joule could make this breakthrough by connecting the dots and why Joule's proposal had taken so long to become accepted. Explanations could be many. At a time when accuracy or quantification was not common practice such as coal was measured in chaldron or cwt, things were based on rules of thumb, often not even necessarily estimates, there would hardly be any credit standing out when Joule shared his stories based on a thermometer that he claimed to be able to tell a difference of 1/200 of a degree (James Prescott Joule). In the end, what one could tell of a one-degree change when industrial processes like coal gas making works in ranges of ten and more degrees or lives were measured by winter and summer? There is, however, a forgotten or easily ignored fact that Mayer and Joule both worked in professions that 1° or even one-tenth of it means a lot to them and makes a lot of difference to their target objects. In their professions, temperature and the change of it is a critical parameter that how Mayer would work carefully to take care of his patients and that how Joule would pay close attention to maintain the right temperature of his brewing process so to make sure his wine a tasty one. Using the human body as an example, it is normal that our human body temperature changes within 1°, ±0.5 °C, during the day and night; imagine what if our body temperature goes up by a degree, we would easily feel the impact and the chances could well be that something might have gone wrong! On the flip side, Joule would probably not be able to succeed in his experiment without his prowess in manipulating temperature. In his experiment published in 1845, to obtain a work of one BTU, that is to cause the temperature of one pound of water to rise by 1°, requires one pound of weight, water or solid body, over the pulley tied by a string to travel vertically 838 feet, close to a height of a 57 story building or structure. To make it practical Joule might have scaled down to a smaller temperature gain so the weight over the pulley travels short enough while his "*thermometer*" would still be able to capture the temperature change with the required accuracy. Of course, Joule could also carry out his experiment by rerunning the pulley a couple of times to make up the needed distance of travel that would cause a measurable change in the temperature of the water. It's a delicate piece of work! Furthermore, Joule was quite confident that his "thermometer" would help him achieve his goal which explains his perseverance after meeting the many silences and rejections before winning his recognition. So did Mayer after meeting rejections to publish his early papers. In the following years, Joule continued his experiments to refine his measurement of the mechanical equivalent heat multiple times and his final number in 1878 became 772.55, the number inscribed on his tombstone after his death on Oct. 11, 1889. In addition, in later years, his work with Lord Kelvin on air or gases, he applied his knowledge of chemistry and Dalton's atomic theory, and established the Joule–Thomson theorem in 1852 which gained wide industrial applications with gas processing and refrigeration etc. So was Mayer in 1871 bestowed with the Rumford Medal of the Royal Society.

Joule's discovery with his delicate experiments convincingly manifested the nature of heat, electricity, motion, and their interchangeability. Heat, motion, and electricity etc. are just different forms of energy; there is no caloric existence for any heat exchange to take place between any of them. This has clearly placed a firm period to the debate over Lavoisier's caloric theory, which was applied by Carnot and many others. Joule's discovery also laid a solid ground for Rudolf Clausius (1822–1888), a German Physicist, Lord Kelvin, and William Rankine to move on to establish the doctrine of thermodynamics; that heat can neither be created nor destroyed and can transform from one form to another became what is the first law of thermodynamics of modern science. The word "*energy*" had since become general use representing all different forms of energy.

More important, Joule's discovery also opened up the windows of possibility for entrepreneurs and engineers to exploit different forms of energy for efficiency gains by reinventing various industrial processes following the Carnot cycle principle. Here comes an entrepreneur and inventor who had been following Joule's progress closely while working around heat to improve the efficiency of a heat engine. In doing so, the entrepreneur accidentally unlocked the value of a coal gas making technology that rendered him great success in steel making.

He is Carl Wilhelm Siemens.

Regenerative Furnace and Metallurgy

Carl Wilhelm Siemens (1823–1883, later Charles William or Sir William), a German engineer, inventor and entrepreneur, was born in a village near Hanover, Germany. He was the fourth son of the Siemens family. At 18, he attended the University of Gottingen and later apprenticed at a well-known engine manufacturer at Magdeburg. Then he launched his career in business development as we it called today and went to London to market an electroplating process that his elder brother, Werner, had developed. In London, he scored success in getting it adopted by a well-known toy maker, Messrs. Elkington. On his next trip, he brought another product of his own invention and also made Birmingham his new home. During the following years, Sir William split his time among marketing electric products with Werner, working at a few shops at different times, and developing his many inventions. The time spent at Birmingham as the center of the ongoing industrial and scientific revolutions in England had proven valuable to Sir William by equipping him with the extensive hands-on experience that helped develop his career in later days. In 1851, he launched his own business to market his inventions, among which a process to move heat around with a regenerative furnace made his fame in steel making and many other industrial processes until his death. Of course, such success didn't come without struggles. It appears that his firm belief in the new scientific discovery and the principle about heat and its nature kept his efforts going.

Since the first locomotive **Rocket** run on the Liverpool–Manchester railway by the British engineer George Stephenson and his son in 1829, railway construction experienced a rapid growth after a slow start. The rail tracks ran from 500 miles in 1838 to 6,600 miles by 1850 and then reached 15,500 miles by 1870 (Stavrianos, 1995). So did the ocean going steamships. Such an expansion called for more steel, a better material, due to the poor experiences with wrought iron as materials for rail tracks, springs, and large metal parts. Wrought iron made rail track was not hard and tensile enough in those working environments. For example, rail tracks in service areas subject to heavy weight or bend lasted only for a few weeks. Steel production at the time had been progressing but slow and operated in small batches with processes such as crucible and puddle furnaces. Set aside its high prices, the production of such steel operations was far behind the market demand. The steel making at the time was definitely in need of a furnace upgrade, a furnace that would be capable of achieving the required enough temperature, high enough to completely melt pig iron. The key lies in finding an effective way to apply heat.

In coal gas making process, getting a heat transferred effectively from a source to a target is slow, time consuming and has some limitations. So is steel making, which also uses a fireplace to heat up a hearth, where pig iron is placed. Coal burning has a limited power to heat up the hearth to a desired high temperature, which is especially true when the scale of the hearth goes up. Here is a paragraph from one of his obituaries presented by the Council of Members telling about Sir William's passion of heat and economizing.

There is no doubt that in his early youth he had studied the theory of heat, and he kept pace with all the later discoveries in regard to it. He had made himself master of the profound investigations of Joule, Mayer, Carnot, and others, and had become well acquainted with the great modern doctrine of the conservation of energy, of which the dynamic theories of heat had furnished such conclusive demonstration. Applying to these theoretical considerations his eminently practical mind, he could not fail to see what an enormous loss of valuable energy was continually going on, by the waste of heat, in almost all manufacturing and industrial processes, and it became his earnest endeavour to discover and introduce means of economizing this wasted power.

Having believed that regenerating the heat lost in the flue gas in a coal fired steam engine system was the way to improve steam engine efficiency Sir William back in 1846 started to apply the scientific principle such as Carnot Cycle and Joule's recent findings to an industrial process heating where coal fired boiler or furnace had been in practice. Such efforts to save coal use in the mid-1800s would certainly attract a broad attention and interest as the industries of iron and steel making, bloom and glass making etc. were all experiencing rapid growth. The demand for coals increased accordingly. Britain's annual production of coal was about 16 million tons in 1815, which increased to 30 million tons in 1830. Between 1853 and 1862, annual average production of coal had reached about 72 million tons already (Historical Coal Data). So did the price of coals, which often made up a big chunk of the cost of products. Product and technology that were thermally efficient would be of great demand. Facing such an opportunity Sir William put his belief and skills into action to develop a more efficient technology or a product so to make the coal heating more efficient.

The time between 1846 and 1856 must have been quite a struggle for Sir William as his attempts to improve the efficiency of a steam engine did not go well as he wished. His first trial in 1847 was to recover the condensing heat from the condenser of a 4 HP steam engine by installing a regenerator at the works of Benjamin Hick and Sons of Bolton in Birmingham. He seems to try to use the regenerated condensing heat to superheat steam. Then two years later he continued his trials at the works of the Fox, Henderson and Co. of Smethwick but success was limited. His concept of generating superheated steam for the steam engine system was, however, recognized by the Society of Arts in 1850. Sir William continued his endeavor in the belief that he would and should be able to make it.

In retrospect, Sir William might have not picked the right working target, the steam engine, even though his endeavor to realize his vision was a right one. In his 1862 presentation to the Institution of Mechanical Engineers in London, he concluded his earlier work with the steam engine as "*many practical difficulties however prevented a realization of the success which theory and experiments appeared to promise.*" As future cases proved he would be better off had he worked on a target with a high or extreme exhaust flue gas wasted to a chimney. Sir William had realized that, in a coal fired steam engine environment, the temperature of heat available to regenerate either from the steam cycle or from coal combustion flue gas was on the low side; any benefits from his regenerator would be marginal. In addition, the steam engines at the time might not have been suited for the use of superheated steam, which could be one of "many practical difficulties." As we know today that steam conditions, either superheated or ultra-superheated, are critical to the performance of an advanced expansion steam turbine system. It was not so back in the 1840s and several decades after because Watt's steam engine was mostly operated with saturate steam at an ambient pressure so that it had a low efficiency at a single digit. Any efforts to improve the efficiency of a steam engine by recovering heat from the low level steam condition would be very limited and marginal. Rather, any potential gain in efficiency would hardly offset the capital of the regenerative system.

In addition, there may also be other factors limiting the benefits of heat regeneration. Sometimes engineering or design comes to play and other times materials adopted or appropriate operating procedures may well determine the fate of how far one's plan could go. In most cases, a combination of these factors often restricts the level of success. Recovering heat from flue gas is not something new; it had been practiced within the circle of ironmasters as early as in the late 1820s. Such a practice had gone unnoticed for a long time, probably, because the ironmasters tended to keep it to themselves. Over the centuries the ironmasters had been working hard to improve the quality of their iron products by achieving a higher temperature, which is a key step. The use of coke invented by Darby in 1709 and the deployment of a strong air blast by using a bellow, driven by an animal or a steam engine, to a furnace or stove where iron is cooked would help achieve a high temperature. Then in 1829, the use of a hot blast by Mr. Neilson was another recent improvement which greatly raised the temperature in the furnace (Cowper, 1860). Such a benefit, however, was restricted by the material used at the time in transferring the heat from exhaust flue gas to blast air. To heat the blast, ironmasters let the blast pass through a cast iron pipe that runs in the exhaust flue before reaching a chimney. The preheated blast then enters the furnace to boost the temperature of the coke firing. Due to the fact that cast iron can't stand a high temperature, however, the preheated blast hardly reaches a temperature higher than 700 °C beyond which the cast iron would lose its strength. In cases like this, material becomes the restricting factor. The steam boiler furnace, however, may not work up for Sir William as the temperature of flue gas would be much lower after giving much of its heat for steam generation. Well, it is unknown if Sir William had tried it but the end game would be the same, at least similar, the impact too marginal.

A few years later in 1856, Sir William's luck started to make its turn when his young brother, Frederick, made a new invention called a regenerative furnace along the line of Sir William's thought of heat recovery, but used a different design and material for heat regeneration. Sir William immediately realized the value of the invention and started to work closely with Frederick to improve it while finding opportunities to deploy the regenerative furnace for industrial application.

Frederick (1826–1904), a German chemical engineer and inventor, went to England to join his brother, Sir William, in 1848 to help his electric business. In the meantime, Frederick worked at the same works as Sir William did previously. In 1856, Frederick filed a Letters Patent (equivalent to a provisional patent) in December and followed in June 2 next year with a full patent titled *"Improved Arrangement of Furnaces which Improvements are Applicable in all Cases where Great Heat is Required"* in Britain. The invention claimed a regenerative furnace plot recovering heat from an exhaust flue at a high to an extreme temperature practiced in many industrial processes such as metallurgy, glass making, and steel making etc. This is about 10 months after the famous patent of blast furnace filed by Mr. Bessemer.

By the 1850s, coal firing or combustion with an atmospheric air blast had routinely been deployed to provide heat to furnaces for metal melting, glass melting, and pottery heating. The scale of the furnaces was small in general and in a batch operation, about 2–4 tons per batch. Typically, achieving a furnace temperature of 2,400 to 2,600 F or higher is necessary to melt the object. However, acquiring such a high heat (temperature) would need not only good quality coals such as hard coals but also skilled labors. These requirements along with the coal fireplace design, on the other hand, would become the restricting factors that hamper the operations going to a large scale. Owners in the fields had been on the look for better ways to improve their operations. Frederick's invention came in at a right time and, with his brother's active introduction, became immediately embraced by the many owners of those works. Even though with a good start, the road of commercialization, however, had taken several turns in the next ten years before seeing a fantastic end.

In a typical operation of an iron or glass melting furnace, the flame from coal combustion heats up a chamber where objects, metal or glass etc. are loaded, then leaves the heating chamber as an exhaust flue typically at a high temperature (containing a significant amount of heat) before entering a chimney. The significant amount of heat in the flue was simply lost into the atmosphere. By applying Frederick's invention to these furnaces Frederick targeted to recover the waste heat from the exhaust flue to heat up the incoming atmospheric air blast; the heated air is then blasted into a fireplace to fire up coal to heat up the chamber where iron or glass are placed to a high temperature. To achieve such a task Frederick adopted the principle of the regenerative system that his brother had been working on but invented a different design to move the heat around. To accommodate the new design (Fig. 6.2), Frederick selected firebricks as structural and heat transfer materials to build the regenerative system, a pair of regenerators (D, D') sitting next to each other as passages alternating between the exhaust flue (F) and the cold air blast (E). The regenerators are placed right beneath the fireplaces (B, B') and the furnace chamber (A) where the target subjects are placed. The firebrick is an excellent heat resistant material, performing well at high temperature environments, and in the meantime retains much more heat, a practice already commonly adopted with coal gas making operations. The design of the regenerator and the number of firebricks deployed would make sure an exit temperature of the exhaust flue below about 200–300 F at the end of the regenerator so to recover most of the waste heat. In an actual operation as shown in Fig. 6.3, the hot exhaust flue leaves the heating chamber to pass through a switch box (functioning as a three way valve) into one regenerator and gives up its heat to firebricks along the zigzag passage before reaching the chimney. Once the firebricks are heated up high enough (retaining enough heat), the hot exhaust flue switches over to heat the other cold regenerator while an atmospheric air blast is routed into the heated regenerator via a second switch box on the chimney side and is gradually heated up to a high temperature close to the temperature of the hot exhaust flue, and then blasts into the coal fireplace. Then the cycle operation repeats itself at regular intervals. By doing so, the hot air will help boost the coal firing intensity to provide better heat to the heating chamber while producing a significant economy of coal by saving nearly 50% relative to the original iron puddling operation. Becoming excited with the potential of this invention, Sir William and his young brother took immediate action to put it to work.

In the early years between 1857 and 1865, Frederick and Sir William applied the new regenerative system to different works with customized designs for heating steel bars, melting iron, copper or glass etc. around Birmingham, Manchester, London, and more places. According to many presentations made to the Institute of Mechanical Engineers by Sir William and the following Q&A discussions as a feedback from owners at the events, the regenerative system worked fine at those works and had resulted in a significant saving of coal use. It, however, did not change the labor intensive operation and the persistent problem of uneven heating within the heating chamber remained existential when the operation moved to large scales due to insufficient heating power of the coal fireplace.

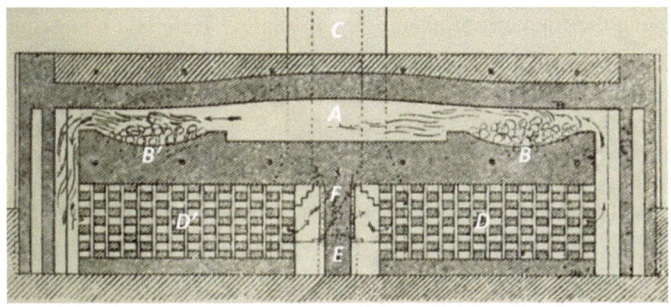

Fig. 6.2 A regenerative system (GB Patent #2861, 1856 by F. Siemens)

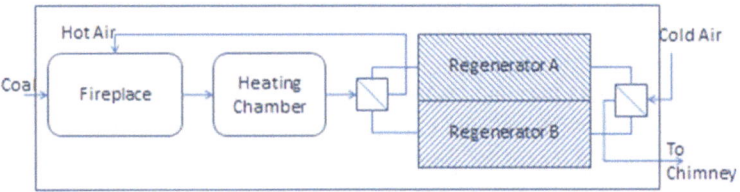

Fig. 6.3 Schematic diagram of the Siemens regenerative furnace

Another problem is that the switch box on the hot exhaust flue side, between the heating chamber and the regeneration system, required frequent maintenance due to the high temperature and the erosive environment. The good news though is that the markets for steel and glass of a good quality were still going strong.

Having tried various designs at different works, one coal fireplace versus two, single heating chamber versus twin heating chambers and different locations with coal fireplaces, Frederick and Sir Williams had finally come to the conclusion that coal firing was the bottleneck and a new type of heat source, gaseous fuel, would help clear the bottleneck. With a gaseous fuel available, the brothers realized that they would be able to take full advantage of their regenerative system by heating up both the air blast and the gaseous fuel, which would further enhance the combustion power to achieve not only a higher temperature but also a more balanced heat distribution within the heating chamber.

Coal gas in the 1860s had already become widely available in cities, towns and even villages in Britain. The Siemens brothers knew that. For industrial use, however, coal gas would be too expensive to make sense, economically. So, what was left available to the brothers? There came another coal gas making process to their rescue—a gas producer with a changed chemistry.

The Power of Integration

On Nov. 29, 1862, the local newspaper *the Glasgow Herald* in Scotland reported about a project deploying the Siemens brothers' new invention. Here is a paragraph from it.

> By this invention, the workmen employed either at glass making or at iron puddling can proceed with their labour with much greater comfort than if they were engaged at a furnace heated in the ordinary way. The furnace to which we refer is the first that has been fitted up in Scotland, and if, as we are informed, the consumption of fuel is much less, and the work better executed, these furnaces are likely to be extensively introduced into glass and iron works in this country.

Confident that a convenient and low cost gaseous fuel was what they would need, the brothers began their hunting for a technology that could deliver it. Soon they found it. More important, their efforts also led to another important development of their invention; that the separation of the fuel gas making unit from the furnace structure would open up a whole range of potential benefits that would unleash the full value of the regenerative system.

The Siemens brothers may not be well aware of the chemistry of a gas producer and the works carried out by Bischof or Ebelmen at the Audicourt Iron works around 1840 that an inflammable gas could be produced when an air blast to a combustion system reduces to a certain level. There seemed that the inflammable gas could be produced by a variety of means. Their initial plan to make a gaseous fuel was to open up a small grate at the bottom of each of the two fireplaces shown in Fig. 6.2 as they might have believed that an inflammable gas would be produced if a flow of air blasts into the coal pile in the fireplace through the grate. Soon, they gave up the idea believing that doing so would not change the labor intensive nature. The brothers then moved on to build a separate gas fuel generator, a Gas Producer, to provide the fuel gas. In this improved design, the Gas Producer becomes a piece of standalone equipment next to the structure, which houses the heating chamber and the regenerative system. The new regenerative system, instead of one pair of regenerators, has two pairs of regenerators, one for preheating gaseous fuel, produced from the Gas Producer, and the other for heating blast air. The preheated gaseous fuel would be introduced through a number of gas inlets into the heating chamber, therein to meet with hot blast air to produce a much more intense heating power. This new design, the integrated regenerative system (Fig. 6.4) had quickly become applied to many glass, iron and steel works around Birmingham and other places across England. What the Glasgow Herald reported on Nov. 29, 1862, was about a project deploying the integrated regenerative system. The operating results from these works were satisfactory by many measures such as a good economy of fuel, an improved quality and yield of the products etc. From the view point of coal gas making, it would be interesting to see how the Gas Producer in this integrated regenerative system was able to unleash a whole range of potential values of Frederick's 1856 invention and how it was able to push the temperature at the heating chamber much higher that would benefit the operations of glass, the iron or steel making. Here are a few of the advantages that the Gas Producer provides relative to the fireplace.

– Preheating both fuel gas and blast air in the regenerator system boosts the heating power to achieve a higher temperature.
– Preheating both fuel gas and blast air saves coal consumption.
– Fuel gas provides a uniform heating, removing the bottleneck of one flame fireplace.

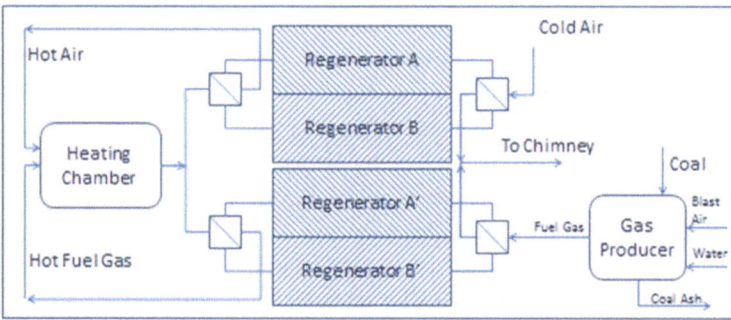

Fig. 6.4 An integrated regenerative system with gas producer

– Fuel gas is much cleaner than the dusty exhaust flue, helping improve the quality of products while minimizing maintenance needs.
– Gas Producer tolerates poor quality coals which are much cheaper than good quality ones resulting in a cheap producer gas.
– Gas Producer could be designed to provide fuel gas to multiple furnaces on site, making the whole operation more economical.

The system (Fig. 6.4) works essentially in a similar way as the earlier version shown in Fig. 6.3 except that the exhaust flue leaving the heating chamber splits into two streams to heat the two separate pairs of regenerators, one for an air blast and the other for the fuel gas manufactured from the Gas Producer. With the Gas Producer separated from the heating chamber structure, the produced fuel gas or producer gas flows via a connecting channel or pipe into the already heated passage of the second pair of Regenerator A' or B', becomes heated and then joined by the hot blast air from Regenerator A or B to fire up the Heating Chamber where the target objects are placed.

Encouraged by the superior heating power unleashed by the Gas Producer, Sir William decided to set up his own shop in order to optimize the technology and develop a system to make better steel at a faster pace. In 1865, he built experimental works in Birmingham to do so and named it the **Siemens Sample-Steel Works** as a platform to develop component technologies and their integrations so to optimize the integrated system as a product that eventually formed the basis of the famous open hearth furnace for the production of steel. In addition, the Siemens brothers also tried to find uses of this new way of heating in other places like the British Plate Glass Works near St. Helen's, in iron and glass works in Germany, iron works in Belgium and in France. The integrated regenerative system was also installed at coal gas works such as the Paris General Gas Works, GLCC's gas works at Brick Lane in London, and probably the Winsor Street Gas Works in Birmingham as well. Among all, steel making appears taking up most of the brothers' effort due to its strong demand and the attractive profit margin at the time (Siemens, 1862).

By 1850, England produced about two million tons of wrought iron annually but only 50,000 tons of steel. The development of railways, bridges and high buildings and ship-building need more steel including structural steel. Due to limited supply, however, steel commanded a good price at the time by selling at 50–60 Pounds a ton while wrought iron at 10–15 Pounds a ton. At the time there was another steel making technology already well under way in its development undertaken by Henry Bessemer (1813–1898), an English inventor and industrialist. By 1862, seven years after his patent, Bessemer had made substantial progress in developing the technology and started to license it to many works for steel making. The Bessemer process takes a molten pig iron from a blast furnace into a converter, a cylindrical steel pot lined with refractory for insulation; then by injecting air from the bottom of the converter to burn out the remaining carbons and other impurities remained in the molten pig iron. The process is called the Bessemer process. By 1867, steel production using the process had reached 500,000 tons annually. Bessemer certainly had his moment. By 1870 the Bessemer process had been widely used in England, delivering a large quantity of steel at a lower price. Now it's the time for the Siemens brothers to catch up.

Leveraging the high heating power made possible by the integrated regenerative system, Sir William was able to take in both pig iron and iron ore directly as feedstock to be heated by the hot fuel gas and hot air blast, which becomes the well-known open hearth process. Although taking a longer cooking time than the Bessemer process it rendered a good quality steel because the longer cooking time makes the control of carbon content and other impurities much easier. In 1868, Sir William moved to a large premise in South Wales to build a larger furnace under his new company, **Landore Siemens Steel Co.** In the following year, he was able to produce 75 tons of steel a week; this number went to 100 tons in 1870. With Sir William's reputation established in the market place, the first commercial open hearth process plant was built at Hallside near Glasgow in 1873. (Barraclough, 1986). This marked the beginning of the Siemens Open Hearth Process that had lasted for more than a century, way outlasting the Bessemer process.

Through his licensee's work in France, in the meantime, another version of the Open Hearth Process using the same integrated regenerative system had been developed in Paris. It's the Siemens-Martin Process, using pig iron and scrap metals as feedstock, which added another competitive edge that contributed to its dominance during the ongoing industrial revolution. The steel from this works rendered Sir William a Grand Price of the Paris Exhibition in 1867. In 1873, the steel produced by using the open hearth processes including the Siemens-Martin Process was 70,000 tons annually, still far behind the 500,000 tons made with the Bessemer process in England. By 1899, the open hearth process had turned the tide by delivering more than 3 million tons of steel against the 1.8 million tons by the Bessemer process.

In summary, Sir William's success lies in his belief in the scientific principles of energy conservation and transformation, his knowledge of chemistry in gas making and his progressive engineering prowess to effectively work around heat energy and acquire the most desired heating power. With increasing deployment of the open hearth process for steel making, Sir William and Frederick continued their efforts to improve and apply the integrated regenerative system and necessary engineering principles to other fields for high quality products such as flint glass, high quality ceramic porcelain, cement and many more in need of clean and high temperature heating; their efforts had been successful. Although they also tested the integrated regenerative system in the Paris General Gas Works and the GLCC's gasworks at Brick Lane as early as in 1862 the commercial deployment at coal gasworks, however, did not happen until about twenty years later when the Dalmarnock Gas Works in Glasgow implemented the integrated regenerative system to improve the performance of its existing coal gas retort benches.

In 1882 during his reign as president of several institutions including the Institute of Mechanical Engineers, Sir William nominated to adopt James Joule's name as one of SI energy units to remember the pioneering work that Joule did to the establishment of thermodynamics that are still in use today, **4,186 J of equivalent work is needed to raise the temperature of 1 kg of water by 1 °C and the equivalent 1 cal equals to 4.186 J.**

Gas Producer and Its Chemistry

The open hearth process as one of the greatest inventions in modern history serves as one of the best examples that demonstrate not only the ingenuity of human engineering but also the power of integration in technology applications. The Siemens brothers had tried many different ways to make their regenerative system work but only found out that their hands were bound in creating a uniform high temperature environment, a temperature high enough to melt pig iron thoroughly, within the heating chamber or furnace. They obviously seemed confident that their trials should be on a right track in terms of the underlining principle about utilizing the potential value of the heat normally lost to a chimney, which may not be so to many others. In retrospect, the Siemens brothers were absolutely correct.

On the other hand, the Gas Producer under the principle of complete gasification invented twenty years ago had been exploited in many applications as well but had found no real value on an industrial scale so far because it produces a gas of low heating value; nor is it suitable for an illuminating purpose. Once the Siemens brothers dusted it off from the shelves and put it next to their regenerative system, however, they suddenly found that the regenerative system would become a much more powerful tool to effectively recover the heat energy in an exhaust flue gas to heat up both the fuel gas and the air blast. In other words, the available producer gas fuel, though having a low heating value, could become a powerful fuel once being heated up in the regenerative system. The recovered heat, once being transferred to the fuel gas, actually increases the potential energy of the fuel gas so to boost it to a more powerful fuel. It is the dual action by recouping more sensible energy from the flues to preheat both the fuel gas and the air blast, which would render more heat in the subsequent combustion to heat up the heating chamber or a furnace to a temperature high enough to thoroughly melt the pig iron to make steel, the most difficult material to melt at the time. It is the producer gas that enables the brothers to scale up an operation capacity conveniently by eliminating the bottleneck previously faced with one flame of coal fireplace; it is such an integration of the Gas Producer with the regenerative system that unleashes not only the value of the regenerative system but also the value of the Gas Producer, which is one of the best engineering principles that have been practiced so far in modern days.

The early design of the Gas Producer used by the Siemens brothers was similar in principle to that invented by Bischof 20 years ago. It is a dry ash discharge system with a square structure of brick and mortar. Coal is fed from the top left of the Gas Producer (Fig. 6.5), and an air blast entering the grate at the lower part into the producer moves countercurrent against the coal bed while reacting with it on its way up. The resulting fuel gas rises through the coal bed and leaves the Gas Producer through an exit at the upper right side of the wall, and then enters into the conduit connecting to the regenerative system of the target applications. There are some additional features as well. The reactor wall on the coal hopper side was designed as an inclined wall supported with a piece of a cast iron plate lined with bricks that provides protection. Such a design would provide additional space holding more coal inside the Producer. The inclined wall connects to an inclined grate where a blast of air enters to fire up the coal bed moving downwards. Another feature is a water trough placed at the bottom of the grate that is in contact with red hot ash. The radiative heating from the red hot ash would make the water inside the trough evaporate; the steam vapor then rises up into the coal bed, which is a feature used by Ebelmen in his invention. The water is automatically supplied to the trough through a pipe connected to a water reservoir.

During normal working status the countercurrent flow in the Gas Producer, coal flowing down against the rising blast air and steam, would result in four distinct zones as described in Chap. 5 that work in sequence to give off the final producer gas while reducing carbon to its completion. The remaining ash is discharged from the bottom. Here is what Sir Siemens provided about the chemistry taking place during the countercurrent flow in the producer.

......the fuel descending slowly on the solid portion B of the inclined plane, Plate 1 (Fig. 6.5), becomes heated and parts with its volatile constituents, the hydrocarbon gases, water, ammonia, and some carbonic acid, which are the same as would be evolved from it in a gas retort. There now remains from 60 to 70 per cent. of purely carbonaceous matter to be disposed of,

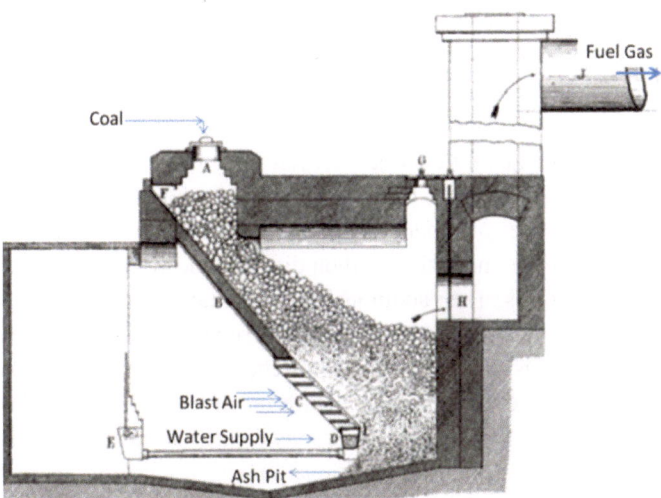

Fig. 6.5 Cross section view of gas producer (Siemens, 1862)

which is accomplished by the slow current of air entering through the grate **C,** producing regular combustion immediately upon the grate; but the carbonic acid thereby produced, having
to pass slowly on through a layer of incandescent fuel from 3 to 4 feet thick, takes up another
equivalent of carbon, and the carbonic oxide thus formed passes off with the other combustible
gases to the furnace.

Obviously, Sir William now had a certain understanding of the chemical reactions taking place in the producer gas making on a macro level which has been applied in the
following century till to the modern day. On a micro level, however, what is happening
right above the grate in the combustion zone between carbon and oxygen, forming carbon dioxide or both carbon dioxide and carbon monoxide, had remained the subject of
many future investigations and studies. Chemists and engineers at the time got it right
that carbon dioxide is the primary gas formed between carbon and oxygen in the air in
the combustion zone and then is reduced to carbon monoxide when passing through the
incandescent zone. This is why their first plan, by retrofitting the fireplace by creating a
grate through which blast air enters into the coal bed to make fuel gas, would not work
because the fireplace would not hold enough coal to form the incandescent zone of 4–5
feet thick to substantiate the necessary chemical reactions giving off producer gas. From a
chemistry perspective, what Sir William described about inside the Gas Producer and the
regenerators looks indeed superficial. Considering the fact that molecular theory was by
large not yet existent, however, it is pretty much how far chemists and engineers at that
time could go in their interpretation of the phenomenon inside the Gas Producer. In addition, Sir Williams also pointed out the occurrence of a continuous reaction taking place
between carbon deposits resulted from the cracking of hydrocarbons and steam present in
the producer to give off additional water gas.

By 1862 the integrated regenerative system had been demonstrated at a number of
works for glass making and iron tube welding in England. The Gas Producers were typically designed to hold about an inventory of ten tons of coal, lasting about five days of
normal operation; that is a consumption rate averaged at two tons of coal per day. The
capability to handle low quality coal had proven a significant value contributing to the
production of a cheap producer gas for industrial use. Each ton of poor quality coal gives
off about 64,000 cubic feet of producer gas that comprises of a wide range of components like hydrogen, carbon monoxide, carbon dioxide, and some hydrocarbons including
entrained tars and light oils. Then additional reactions take place while the producer gas
passes through the regenerator, resulting in more permanent inflammable gases.

When the producer gas passes through the Regenerator and becomes heated up to a
temperature of close to 1,600 °C, a temperature much higher than that in the Gas Producer, components especially those hydrocarbons formed at much lower temperatures in
the producer gas would be subject to additional chemical changes due to further cracking of the entrained hydrocarbons, light oil and tars in the passage; and water vapor
therein would react with the carbon deposit from the cracking to give off more permanent inflammable gases. In the end, the 64,000 cubic feet of producer gas per ton of coal

would become 72,000 cubic feet of fuel gas, a 12.5% increase. When such a fuel gas and hot air meet at the inlets to the furnace heating chamber for glass or iron making the combustion would release a much more intense heat so to raise the temperature way above the boiling temperature of pure iron, about 1571 °C, a temperature good enough to produce a better quality of steel and glass products by driving out more impurities from their precursors, noting that there is an exception for iron ores containing a high phosphorous element at the time. In addition, the producer gas makes the scale-up possible by arranging more burners around the furnace chamber so to make heat distributed to where it needs to be, achieving a uniform heating effect, which a traditional one flame fireplace was not capable of.

In short, Frederick's early regenerative system, once integrated with or enabled by a gas producer, becomes a much more powerful tool that made the Open Hearth Process compete well with the Bessemer process in steel making.

Producer Gas and Regenerative Retort

Once having established the Open Hearth process for steel making and high quality glass making the Siemens brothers then turned their attention back to coal gas making around 1880. This coincided with the time when electric lighting such as the arc lamps came to market place; and a new incandescent light bulb that Thomas Edison invented in 1879 did attract a lot of attention. As early as in the 1850s the arc lights powered by batteries had actually started to light up many lighthouses and some open places in Europe. The threat to the coal gas business seems real but very limited because the arc lighting, a stunningly intense white color, is only suitable in open spaces but not in a housed spaces. Nonetheless, it did spur the coal gas business to start to shift to other fields such as heating of commercial buildings and domestic houses, and cooking etc. Around the time in 1856 the advent of the Bunsen burner invented by Robert Bunsen (1811–1899), a German chemist, to some extent, helped sustain the coal gas business a bit longer. Not like the Argand burner in broad use, Bunsen designed his burner by premixing air with fuel gas before reaching the burner tip or opening to give off illumination; therefore, the new way of combustion of coal gas becomes more efficient and cleaner without smoke, which enhanced the coal gas lighting experience more pleasant that helped sustain the coal gas for lighting business. When incandescent lighting arrived, even though still expensive at the time, the landscape of the burgeoning coal gas lighting business had changed; the new lighting with a much brighter incandescent glow could be turned on and off by simply flipping a switch at a fingertip had a great appealing to many customers.

Feeling the pinch, many coal gas companies were forced to drop their coal gas price in order to stay competitive. While simply dropping the price of coal gas did help coal gas companies as a short term tactic cutting into profit margin would hardly be the solution in the long run. Coal gas companies, therefore, started to look for ways to reduce their operating costs for coal gas making by improving and upgrading coal gas making process while in the meantime developing new market for coal gas consumption; residential heating and cooking and industrial heating are a few examples in the direction. Products such as gas ovens, gas cooking tops, water heaters, etc. had been invented and picked up by coal gas companies to add to their equipment rental list. In Shanghai, for example, electric lighting first appeared in August 1881. Responding to such a threat the British Concession Gas Works had to drastically reduce its coal gas price to 3.5 Silver Dollars for every thousand cubic feet of coal gas in 1881 noticing that coal gas commanded a price of 4.5 Silver Dollars per thousand cubic feet in 1865. In 1882 coal gas prices dropped further to 2.5 Silver Dollars in order to retain the customer base. To stay competitive, the British Concession Gas Works also decided to improve the quality of its coal gas and building more storage space, which had taken a much longer time to realize. In 1879, Shanghai Concession Gas Works introduced its first cooking top for customers to try out; it also imported small gas engines of different sizes between 2 and 8 HP in 1882 for commercial use such as in printing houses (City, 2023). In England, gas companies started to work harder to continue to grow their gas business. Prior to 1880, coal gas companies enjoyed a time of a fast and, in some cases, violent growth through erecting new gas works, expansion or amalgamation and so forth. In London, GLCC had increased its coal gas capacity to 11 billion cubic feet in 1877, about 65% of all coal gas manufactured in the city after absorbing other gas companies, *the Imperial and Independent*, *the City of London*, *the Great Central*, *the Equitable*, and *the Western* (Dresser, 1877–1878). In New York City, according to Con Edition's website, there were six gas companies in 1876, *the New York Gas Works* first built in 1823, *the Manhattan Gas Light Co.* in 1833, *the Harlem Gas light Co.* in 1855, *the Metropolitan Gas Light Co.* in 1858, and *the New York Mutual Gas Light Co.* and *the Municipal Gas Light Co.* by 1876. According to city's charter, each of the companies should serve coal gas to customers in its own territory. Digging street to lay more gas mains then became a way to compete for more customers so that in 1880 city's major companies had to come to an agreement with a fixed price of coal gas in order to end the constant digging on streets, a business practice obviously not legal in today's environment.

From a technology view point, the year 1880 seems a turning point for coal gas making prior to which it remained essentially stagnant, having been using coal fireplaces to heat up retorts, horizontal or inclined, in a bench structure to a required temperature and hold for some time to maximize the coal gas releasing from coal. It is in a cycle operation and each cycle lasts about six to eight hours or longer depending on the coal used and the final retort temperature. At end of each cycle, coke needs to be dumped out and fresh coal to be replenished; it is a labor intensive and harsh working environment. Although some measures had been taken to make the work easier such as using mechanical tools to pull out coke and replenish coal etc. coal fired retort operation remained a primitive process that requires a significant amount of coal to heat up the retorts, which is polluting to the local environment as more retort facilities had been built into the neighborhoods.

After 1880, however, the coal gas market started to shift primarily due to the emerging competition from electric incandescent lighting and the changing coal market as well. Coal gas companies were compelled to look for options to improve their coal gas making experience. Overall, the timing could not be better when the Siemens brothers made their comeback to find their renewed interest in coal gas making with a new design of the gas producer. Instead of a squared brick–mortar design the Siemens brothers brought up a cylindrical vessel made from wrought iron and inside lined with firebrick, a design that had already been practiced for years in America. The Siemens brothers' improved integrated regenerative system caught the immediate attention of Mr. W. Foulis who was the General Manager of the Glasgow Gas Corp. Trust, owning several gasworks in the Glasgow area. The Siemens brothers had reached an agreement with Mr. Foulis, most likely around 1880 and 1881, to test out the Integrated Regenerative System at his Dalmarnock Gas Works, sitting on the most eastern side of the city. Soon the Siemens brother scored another success with their new design (SAS, 1882–1883).

The Dalmarnock Gas Work was the largest gas work in the region, holding 750 retorts. Mr. Foulis wanted the Siemens brothers to test their new regenerative system first on four sets of seven retort benches; each of the retorts has a dimension of $9' \times 18'' \times 13''$. To apply a heat regenerative system to a gas fired retort works the typical layout would be a raised structure where the retort sits on top and the heat regenerative system right beneath the retort house, which would conveniently allow hot flue gas from the retort house directly flow downward into the heat regenerator **A** and **B** alternately and therein give up its heat to heat up the regenerator before reaching chimney (Fig. 6.6). The gas producer would be located right next to the retort structure, either front or back, so that hot producer gas would enter immediately to the combustion chamber where it meets preheated air from the regenerator **A** or **B**. These measures are designed to minimize heat loss. When it came to the Dalmarnock Gas Works, however, it was an existing retort house; raising up the retort house high would have to rebuild the work completely, which was impractical. Siemens brothers took a different approach by going under, placing the heat regenerator right under the retort house, by digging about ten feet into the ground for the necessary spaces for the regenerators and the producer, which the latter stands right next to the regenerators on front side of the retort house because the hot coke, once discharged from the retorts, could be guided directly to the producer as feedstock, different from previously using low quality coal. The Scotch cannel coal used in the gasworks resulted in a poor quality coke of little value, a convenient reason to dump it to the producer next to it, when appropriate. Once hot producer gas and hot air meet at the combustion chamber right beneath the retort in the center area, the gaseous flame would spread out evenly around the retorts inside, creating a much better heating environment than the one flame of a coal fireplace.

To make sure the test successful, the Siemens brothers placed one small gas producer ($3'$ ID and $7'6''$ H) to each of the four sets of retort benches. The cylindrical producer, with a wrought iron shell lined with firebricks inside, receives hot coke appropriately

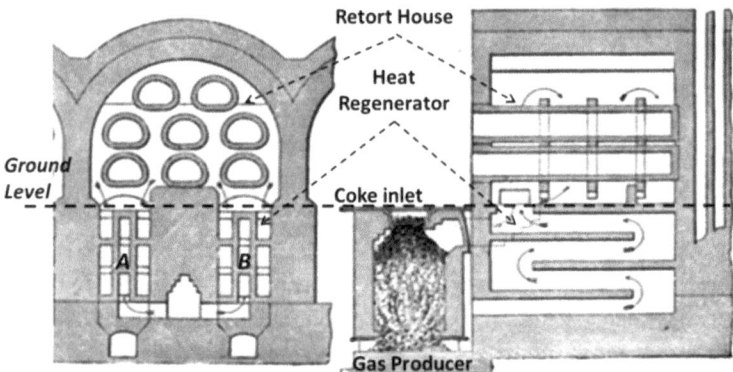

Fig. 6.6 Gas fired retort bench with heat regenerative system (SAS, 1882–1883)

from the top opening while producer gas passes out at the top right corner directly to the center chamber for combustion. After upgrading the direct coal fireplace with producer gas, each of the seven retort benches had been expanded to house eight retorts because of the extra space created after removing the coal fireplace. The upgrade, after some trials and modifications, proved successful not only by creating an improved heating experience but also reduced fuel consumption by more than 50%. Previously, heating a seven retort bench consumed 65–75% of discharged coke; after the retrofit, heating eight retort bench reduced coke use to somewhere between 30 and 50% of the previous rate noting that each retort was now loaded with 38% more coal. The upgrade did result in significant benefits to the Dalmarnock Gas Works with fuel savings and onsite destruction of the low value coke while manufacturing more coal gas due to the improved heating with the integrated regenerative system.

Mr. Foulis was pleased with such a success that *"The new system of firing being made so simple that there was scarcely any possibility of failure likely to arise in ordinary practice if it was superintended with but a moderate amount of care."*

With the first four sets of seven bench retort retrofitted successfully the integrated regenerative system along with the lessons learned from the trials was applied to the remaining eight sets of seven retort bench, and then followed by additional 22 sets of seven retort bench. Soon, the integrated regenerative system found its value and many other coal gas works such as the Winsor Street Gas Works, and the Tipton Gas Works near Birmingham and other countries beyond Britain followed the suit until the vertical retort technology would then emerge in the early twentieth century.

The regenerative retort process did not arrive at China until 1900. Around the time the Shanghai Gas Co. replaced two sets of its 11 five retort bench with the integrated regenerative horizontal retort system. The technology upgrade enabled the coal gasworks to extract 36% more coal gas from each ton of coal consumed at the works. Based on available information it appears that the integrated regenerative system eventually upgraded all remaining old retort benches in the next few years, manufacturing more coal gas to meet the increased demand for coal gas in the city. The upgrades remained in operation until around 1920 when the Shanghai Gas Co. introduced vertical retorts into its gasworks, and eventually phased out all the regenerative horizontal retorts.

In America, the story is different. The deployment of the integrated regenerative system to coal gasworks seems limited as America had started to take on a different technology that would revolutionize the coal gas making experience.

References

Barraclough, K. C. (1986). *The development of the early steel making process.* Ph. D. Thesis, the University of Sheffield, Sheffield

City, S. (2023). *Shanghai coal gas.* Retrieved 2023, from Shanghai City Chronicles: http://www.sht ong.gov.cn/Newsite/node2/node2245/node4516/node55027/index.html

Cowper, E. A. (1860). On some regenerative hot-blast stoves working at a temperature of 1300 fahrenheit. In *Proceedings of Institution of Mechanical Engineers* (p54). Institution of Mechanical Engineers.

Dresser, C. (1877–1878). London gas works. In *Proceedings of American gas association* (p. 162). American Gas Assosiation.

Mendoza, E. (2023). *Sadi Carnot, French Engineer and Physicist.* Retrieved May 2023, from Encyclopedia Britannica: https://www.britannica.com/biography/Sadi-Carnot-French-scientist

Morse, E. W. (2018). *Thomas Young.* Retrieved Aug 2022, from Encyclopedia.com: https://www.enc yclopedia.com/people/science-and-technology/physics-biographies/thomas-young

SAS. (1882–1883). The heat regenerative system of fired gas retort description. *Scientific American Supplement.*

Siemens, C. W. (1862). On a regenerative gas furnace, as applied to glasshouse, puddling, heating. In *Proceedings of the Institute of Mechanical Engineers* (p. 21). Institute of Mechanical Engineers.

Stavrianos, L. S. (1995). *The world since 1500.* New Jersey: Simon & Schuster Co.

Wikimedia Commons. (2006). *File: Joule's heat apparatus.JPG.* Retrieved June 2023, from Wikimedia Commons: https://commons.wikimedia.org/wiki/File:Joule%27s_heat_apparatus.JPG

The Alternative Illuminating Gas

7

Coal gas making in America started late than in England, and coal gas making technologies deployed in America prior to the 1870s had been in general of the British origin. With coal gas market picked up speed from the 1850s America eventually overtook Britain as the largest coal gas market. Felt the pressure from the emerging electric lighting the drawbacks of the traditional coal gas making process by retort had become more obvious than ever before. Industrial owners began to seek alternatives, technologies more competitive, clean, and convenient. Although the Siemens Brothers' Gas Producer was capable of converting completely coal into gas which had been used to heat up many industrial processes including coal gas making its low heating value, unfortunately, prevented it from being used for lighting purposes. There came a timely invention brought up by Thaddeus Lowe, the water gas process in the early 1870s, which soon picked up the market of gas lighting. Noticeably, such an invention took place on the other side of the Atlantic Ocean, the United States of America. The new and unique technology also made America a prominent player in the coal gas making business from the 1870s in a much broader way, which also laid a solid foundation not only for the coming synthetic chemical industry but also led to the development of the modern gasification technology as of this day.

Coal Gas, Hydrogen and Ballooning

Since Lavoisier established the existence of the gaseous state many venturous and curious individuals started to exploit the use of the lighter-than-air gases to create a lifting power for aerial flights. Gases such as hydrogen discovered by Cavendish and the water gas by Fontana were utilized by retaining the manufactured gases in a balloon to make the

© The Author(s), under exclusive license to Springer Nature Switzerland AG 2024 77
Q. Zhuang, *From Coal to Hydrogen*, Synthesis Lectures on Chemical Engineering and Biochemical Engineering, https://doi.org/10.1007/978-3-031-55586-2_7

flight possible because the density difference of hydrogen or a water gas from air would create a lifting power, which is determined by the type of gas and the volume of it. Such knowledge had triggered many imaginations on what ballooning could do by flying up high to explore the unknown "heaven", which aroused much public interest. During 1794 and 1799, after the outbreak of the French Revolution, the French Army used balloons for aerial reconnaissance to spot on the positions and moves of the enemies during the campaigns of the Fleurus, the Liege and the Brussels against other European rivals. It appears that the French Army used a gas furnace to manufacture coal gas by decomposing water over red hot charcoal. Although reacting sulfuric acid with iron would be relatively easier to generate hydrogen, a more effective gas than the water gas the supply of sulfuric acid was banned at the time for any uses other than the making of ammunitions, which is more critical for the war (Haydon, 2000) (French Aerostatic Corps).

This is about the time when Murdock started to play around with coal gas he made himself. A little more than sixty years in the North America, a time when coal gas had become available, Lowe used the coal gas to lift up his balloons in his venture and more.

Thaddeaus Lowe (August 20, 1832–1904), a self-made American chemist, aeronaut, inventor and entrepreneur, was the second child of five born into a wealthy family in Jefferson Mills, New Hampshire. At age of 26, Lowe had already become a well-known balloon builder and aeronaut, and had an ambition to cross the Atlantic with his own balloon. In 1859, Lowe built a gigantic balloon the he named the *City of New York*, which stood 200 ft tall with a lifeboat beneath the observation basket of 130 ft in diameter, and could hold 725,000 ft^3 of lifting gas. For his Atlantic crossing, Lowe selected the site of the Crystal Palace, New York City to launch his test flight on November 1, 1859. Although the Crystal Palace, built to house the first World Fair in North America in 1853, was destroyed in a fire the year earlier the open ground located in midtown Manhattan is not only an ideal place for publicity but also conveniently close to the coal gas works of the New York Gas Light Co. on the Grand Street. The test flight, unfortunately, did go as planned due to a hiccup in the communication with the superintendent of the gas works.

Upon inflation began Lowe found out that the coal gas works could only provide 50,000 ft^3 per day instead of per an hour, which would take at least fifteen days to fully inflate his gigantic balloon. By then, weather changes might well be against his mission (Evans, 2002). Facing the buoyant public interests Lowe had finally to abort his plan. Then in the next few years, Lowe made several more attempts from Philadelphia with coal gas provided by the Philadelphia Gas Work at Point Breeze, which at the time had a capacity of 300+ million ft^3 of coal gas daily, large enough to provide extra lifting gas to Lowe's balloon. More delays and failures forced Lowe to revamp his plan and instead to do more test flights in order to learn more about the air flow and weather related air dynamics over the Atlantic seaboard. Then, the breakout of the Civil War not only abruptly ended his venture but also made him the first prisoner of war. When his balloon *Enterprise*, one of his old balloons, was filled up with coal gas from the Cincinnati Gas, Light & Coke Co. and took off from Cincinnati, Ohio on the early morning of April 19, 1861, instead of heading east, the wind carried him and his *Enterprise* south/south east and landed in South Carolina, now controlled by the Confederate Army. After proving himself a civilian and a balloon adventurist, instead of being a spy for the Union, Lowe finally returned home (America, 1907). Soon after, Lowe joined the Union Army and became the pioneer of the US military aerial reconnaissance during the war.

At the breakout of the war, industrial activities in the US had in general concentrated in the Union states, which are reflected with the development of coal gas works in those states. Cities like Baltimore, Philadelphia, New York, Boston, and Cincinnati etc. had built their own coal gasworks. It appears that President Lincoln had considered the possibility to help end the war early by using the ballooning technique to collect information on the Confederate Army. However, he would need a man with the right expertise and skills to work on the possibility. Recommended by Joseph Henry, Secretary of the Smithsonian Institute, Lowe was called up to Washington DC to demonstrate his ballooning. Lowe made it to DC in early June 1861 and was received by President Lincoln on the 11th. The demonstration took place at the Columbia Armory on the 18th. The Columbia Armory, currently occupied by the Smithsonian National Air and Space Museum, is well picked as it was right next to the Washington Gas Light Co. established in 1852, currently occupied by the National Museum of the American Indian. Lowe rose up with his *Enterprise* to 500 ft. high above the Columbia Armory and dispatched a telegram to the White House. The telegraph reads *"To President United States: This point of observation commands an area nearly fifty miles in diameter. The city with its girdle of encampments presents a superb scene.....T.S.C. Lowe"* (US Army Balloon Corps).

Under a civilian contract, Lowe was made the chief of aeronaut in charge to establish the US Army Balloon Corps, under the Topographical Engineers established in 1838, in anticipation of providing field aerial reconnaissance for the Union Army. With allocated resources, Lowe was quick in action and jumpstarted to recruit a team and build balloons and necessary equipment for the anticipation.

To be effective on the battlefield, Lowe understood that he had to build the balloons robust, mobile, and capable to act as swiftly as possible. So Lowe designed three small balloons (*Eagle, Constitution and Washington*) for one person mission and four large ones (*Union, Intrepid, Excelsior and United States*) for heavy loads. From his last experience at Cincinnati, Lowe also directed all the balloons to be tethered through ropes held by members on the ground, which otherwise could drift away from the unpredictable air dynamics. Of importance, however, he would need a portable gas supply available at reconnaissance stations. After several trials, filling balloons with coal gas from the gasworks in DC or other cities proved simply impractical because the many trellises bridges built on rivers and other hazardous conditions such as telegraph wires and tree lines between coal gasworks in DC and the reconnaissance stations made the towing of the balloons an enormously challenging. In one of the early missions, Lowe's team spent a whole night towing the *Union* to traverse a 3 mile distance from the coal gasworks in DC to the reconnaissance station on the other side of the Potomac River.

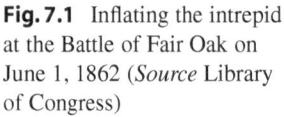

Fig. 7.1 Inflating the intrepid at the Battle of Fair Oak on June 1, 1862 (*Source* Library of Congress)

To contrive a portable gas generator, Lowe realized that the gas generation system had to be mobile and robust for operation and quick to generate enough gas to inflate balloons in a timely fashion; and more, it had to be compact enough to traverse the battlefield that might well be on rugged terrains or hills. Fully aware of the well-established coal gas making technology Lowe, nonetheless, turned to the old hydrogen making by reacting iron with dilute sulfuric acid, which he believed to better serve his needs. Sulfuric acid, though expensive, was easily available from the industrialized north. Here, Lowe exhibited his chemistry knowledge, creativity, and the mindset of designing to purpose. Once he had all the parts and pans put together, the final product, named the Lowe Hydrogen Generator, would give the old art a totally new life, novel, compact, quick-to-action and robust (Fig. 7.1).

Lowe placed the generator, an enforced and acid proof wooden box partially filled with iron turnings and water, firmly on the frame of a normal army wagon. The generator has two openings on its top, one for pouring sulfuric acid and the other connected air tight with a rubber pipe as hydrogen passage out. To make hydrogen operators would follow Lowe's procedure by pouring carefully sulfuric acid, in several batches, through a funnel to the iron turnings sitting in a water pool, and hydrogen would rush out of the box into cleaning boxes to remove potential foreign residues, and then a control valve before entering balloon (Evans, 2002).

Here is how the system works. Each balloon was equipped with two wagons and each wagon is drawn by four horses. A spare wagon drawn by two horses held 4×800 pound of iron turnings and 10×4 gallon of sulfuric acid plus spare rubber parts. Test indicated that one generator could fill up a balloon in three hours and fifteen minutes. If urgent, running two generators would cut it by half. The Lowe Hydrogen Generator was put into services about October 1861 and "worked admirably", as Lowe reported afterword (US Army Balloon Corps), in inflating the **Washington**. Then, between October 1861 and summer of 1863 the Balloon Corps provided field generals with reconnaissance information at several battlefields. For example at the Battle of Fair Oaks in June, 1862 two generator wagons standing next to each other were deployed at the same time to fill up the **Intrepid**

(Fig. 7.1), which was also expedited by transferring hydrogen gas from other balloons. Lowe's invention, different from all previous hydrogen generating equipment adopted by many other aeronauts, worked like a charm and his then reconnaissance had aided General McClellan during the battle with information about the Confederate camps and its moves (Lowe, 1911).

It appears that Lowe attempted to obtain a patent of the hydrogen generating system but failed to do so through the military system. Lowe's invention using hydrogen was certainly against the "norm" of the time. However, it did serve the needs well and made him the pioneer of aerial reconnaissance in US history. From this point on, Lowe further exhibited his personal traits and expertise in many more inventions including the development of another unique coal gas making process that would replace the old retort process for coal gas manufacture, which also made him a good fortune.

Lowe and Water Gas

Lowe's service with the Union Army ended early with his resignation in May 1863. Then, the Balloon Corps also ceased to exist three months later. Lowe returned to his home in Philadelphia and then moved his family to Chester County near Phoenixville (Auge, 1879). There, Lowe resumed his academic enthusiasm and innovations in areas with strong market demands. In the meantime he also made his consulting available to those who were interested in ballooning as balloon aeronautics were gaining popularity in many counties like Britain, Germany, and Brazil etc. One of his inventions, for example, is an ice making machine by using carbon dioxide as coolant, which he also tried to turn it into a business for beef transportation between Texas and New York City (Lowe, 1867). In the meantime, Lowe must have noticed the fast growing market in gas heating, illuminating, and the challenges that the old coal gas making via retort had faced, at least, from his experience with ballooning. After moving his family to a house in northern Norristown, Pennsylvania in 1871 Lowe stepped up his efforts to investigate a better way for coal gas manufacturing. Like what Murdock did in 1792, Lowe also used the gas manufactured from his apparatus to light up his Norristown property (Auge, 1879). The house has become a landmark nowadays in Norristown for visitors.

Since Mr. Edwin Drake drilled his first oil well in Titusville, western Pennsylvania in August 1859, the production of oil from the area had increased to 5.4 million barrels in 1870 from 0.2 million barrels in 1860; it reached a peak about 1882. Unlike today, the oil back then had really not much use except that it was used simply to produce lamp oil or kerosene, a mixture of paraffins, naphthenes and aromatics containing 11–15 carbons depending on its source, which was generally used for illuminating in rural areas and places where had no access to coal gas. Other than kerosene, the rest of the oil, naphtha and heavy oils were of little commercial value. It appears that Lowe was quick enough to look into this 'waste oils' for manufacturing illuminating gas. In August 1872, Lowe filed several patents of an apparatus with the US Patent office (US Patent# 130,381/382/383) to crack naphtha and heavy oils to make gas fuels for illuminating and heating. In 1873 he erected a gas making plant in his previous neighborhood, Phoenixville, which was most likely of the nature by cracking the waste oils. There were several more followed with technologies of his later inventions. In 1875 Low established a gas company, the People's Fuel, Gas and Light Company, in Norristown to manufacture and distribute the fuel gas for illuminating and heating for the local community but such an effort did go far enough. Built on these experiences Lowe turned his focus back to coal.

In March of the same year, Lowe filed a patent (US patent #167,847) on a process using coal to manufacture coal gas for illuminating and heating uses. This invention had become the bedrock of the future water gas processes utilizing coals as feedstock. It soon became widely adopted in the market places both in the US and worldwide, not only extending the life time of coal gas for illuminating by replacing the retort process but also enabling the onset of the modern synthetic chemistry for artificial fertilizers, chemicals and liquid fuels, which would take place during the early part of the twentieth century. Lowe's design of the new gasifier is almost identical to the gas producers in principle, but different in how it is operated. Lowe's design splits the operation of a producer into two cycles, Blast and Run, working alternately. The Blast cycle is to stoke heat inside the Gas Generator and the Superheater and to preheat air at the Air Preheater utilizing the flue gas leaving the Superheater; and the Run cycle to generate a blue water gas with superheated steam generated at the Superheater or a carburetted water gas if impregnated with naphtha spraying (Fig. 7.2). In the Blast cycle, the Gas Generator acts essentially as a gas producer of coal reacting with air, and as a blue water gas generator in the Run cycle by reacting red hot coal with superheated steam.

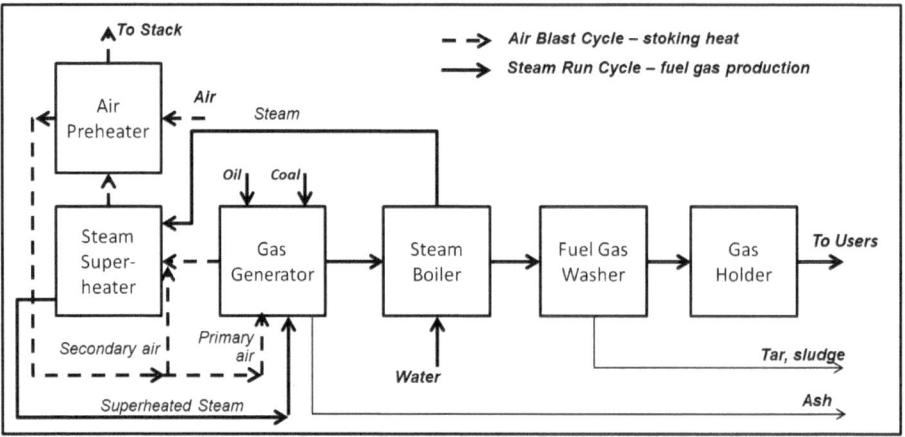

Fig. 7.2 BWG/CWG schematic flow Basis Lowe's 1875 US Patent #167.847

The advantages of such a process are obvious. Different from the gas producer that the Siemens brothers put into use in steel or glass makings, the Lowe process in the Run cycle produces a coal gas having a much higher heating power due to its high concentrations of hydrogen and carbon monoxide with a small amount of nitrogen due to air leakage around the Gas Generator (Table 4.1 III & IV). Without oil spray from the top of the Gas Generator, it would produce a straight blue water gas (BWG) as the gas gives off blue flame upon burning. If the blue water gas is impregnated with cracked oil gas from sprayed oil the resulting illuminating gas would become a carburetted water gas (CWG), possessing a much more candle power than the coal gas manufactured from a retort operation. In addition, the Lowe water gas process can easily produce a large quantity of coal gas because of its complete conversion of coal, making it more cost effective as an alternative illuminating gas.

Interestingly, from an innovation perspective, each component in the Lowe process appears familiar and had been practiced, here and there, in principle. For example, the Gas Generator and the Steam Superheater are similar to what the Siemens brothers had invented in their Regenerative System for the open hearth process; injecting steam on to hot coal or carbon had been known and attempted for decades after being discovered by Fontana in 1780; the Air Preheater and the Steam Boiler already adopted in blast furnace operation previously. All these facts, however, have not diminished in any way the value of Lowe's invention as future industrial experience has proven it. The unique way that Lowe pieces these components together into his grand scheme has proven novel, well integrated, and efficient. The Blast cycle operation makes the heat-up of the coal bed both effectively and efficiently so that in the Run cycle the red hot coal reacts with steam to produce a inflammable coal gas comprising 80–90% of hydrogen and carbon monoxide, a result that could only be achieved with pure oxygen as oxidant, remembering that air separation technology was not invented until the end of the century. The resulting BWG has a heating power more than twice that of a producer gas, and once combined with cracked oil gas, would become a powerful illuminating gas, CWG. The combined result of this process design enabled with the unique operational philosophy has delivered a strikingly new product, demonstrating Lowe's ingenuity, hands-on and design to purpose skills that he had exhibited previously in developing the hydrogen generator during the Civil War. Speaking of this, it is worth mentioning another relevant encounter that happened back then.

Before Lowe was made the chief of aeronaut of the Balloon Corps the Union Army had actually been working with John Wise (1808–1879), American Aeronaut, who was potentially poised for the post. Had Wise succeeded in accomplishing the reconnaissance run before the Battle of Bull Run in the summer of 1861, Lowe might have worked for Wise in the Balloon Corps. Professor John Wise, as one of the pioneers of American aeronautics, had been recognized since his maiden balloon flight from Philadelphia on May 2, 1835 (Evans, 2002). He was deemed one of the few experts on aeronautics in America and authored a treatise of "*A System of Aeronautics*" in 1850, which he later updated it in 1873. There seems no surprise at all that Wise had received support early on from some officials of the Union Army including the Topographical Engineers. To make the reconnaissance effective, Wise also proposed a portable gas generator for field service. Wise's proposal, however, called for a high temperature process by reacting charcoal with steam, the principle deployed by the French Army in 1793. To make it portable, Wise proposed two identical reactor of $18''$ diameter acting like retort, each filled with charcoal and iron turning and placed in a boiler compartment, which would be heated with a wood fireplace underneath. The set of compartments and fireplaces was mounted on an army wagon frame. The required steam would be generated from another wood fired boiler placed on a second wagon. The generated steam is connected to both reactors with a split pipe having a stop cock on each of the branches to regulate and shutoff the flow of steam. In preparing to make gas, the fireplaces heat up, respectively, the compartments (the reactors) to white-hot, and in the meantime, the steam boiler to make steam ready. Then, open one steam line to let steam into one of the reactors. Once inside the reactor the steam would react with the red hot charcoal to give off water gas similarly containing hydrogen, carbon monoxide and a small amount of carbon dioxide. When the temperature in the reactor drops low that the reactions become insignificant, the steam would be switched to the other reactor so the gas making would continue. Such an alternation would generate a continuous flow to fill a balloon. Wise estimated that his system should produce 5000 ft^3/h of water gas, which would inflate a 20,000 ft^3 balloon in 4 h.

The Topographical Engineers requested experts in Philadelphia, the Morris, Tasker and Co. and Dr. John C. Cresson, president of the Franklin Institute and president of the board of the Philadelphia Gas Works at Point Breeze, to review Wise's proposal (Haydon, 2000). The review, however, suggested that Wise's plan would be inadequate to support his estimates. Then, the revised plan would become too costive so that the Topographical Engineers declined Wise's proposal.

From the perspective of gasification what Wise had proposed would probably work in principle at a high temperature, however, had not yet been proven commercially. In addition, a few other technical details such as if the iron turnings would add any value and if the portable system, if made, would work to serve its purpose and so on would subject to questions. What was intriguing though is that the core principle in Wise's proposal, first heating up charcoal and then blowing steam on to the white hot charcoal in alternation, had been reflected in operating the water gas generator of Lowe's invention. Lowe might have known Wise's proposal of a high temperature system, but decided, nevertheless, to move forward with a low temperature one of a different chemical discipline to build his portable hydrogen generator system, which proves Lowe "*at an early age developed his marked aptitude for applied science*" (Haydon, 2000). Lowe's approach to problem and his engineering philosophy to design it to purpose were on a sound foundation that can only be acquired through hands-on experience, knowledge, and creativity.

To demonstrate his new invention to make water gas, Lowe did what was necessary by investing to erect an experimental facility under the Lowe Manufacture Co. in the lower part of Norristown to investigate and improve water gas system of his invention (Auge, 1879).

The Lowe process represents a leap in technology for coal gas making, delivering not only the flexibility and good quality of coal gases, the BWG for domestic heating and cooking and the CWG as an excellent illuminating gas, once impregnated by oil cracking. Although Lowe's patent claims that his process could use different coals and biomass, high rank coals such as anthracite or coke tend to work better for the process, which had been the case for most of the future gas works. However, his design philosophy to integrate the gas producer, invented by Bischof and Ebelmen about 40 years ago, for efficient internal heating of both the Gas Generator and the Superheater, and the subsequent heat recovery for generating steam and preheating the blast air had been well executed by the unique way of operating it in a Blast and Run cycle to deliver a high quality water gas. The Lowe process is novel, effective, and efficient in rendering such a quality and low cost BWG and CWG. From engineering perspective, Lowe's design of equipment such as generator, superheater, scrubber and condenser etc. had departed the conventional materials of brick and mortar by adopting cylindrical steel plate lined with refractory materials to minimize heat loss. This marks an early form of modern process engineering. During commercialization in the following decade, Lowe further improved the Lowe water gas process by making it a compact product that would achieve a wide adoption worldwide.

In 1886, Franklin Institute doled out a Silver Medal of Honor to Lowe's 1875 invention in the spirit of honoring invention and discovery in the fields of science and art that "*....contributed most largely to the welfare of mankind*" based on extensive investigation and assessment carried out by a specially committee (Institute, 1886). The investigation provided some insights of the experimental coal gas works developed by the Lowe Manufacture Co.

> It (the experimental works) embraced a gas works, complete for the generation of water gas (BWG), having a capacity of 5000 ft^3 per h.......In a suite of three rooms, occupied by this company, a very attractive display was made of the applicability of water gas for domestic lighting and heating. Its display embraced an open fireplace of the usual pattern, various forms of fixtures adapted for a special system of incandescent lighting (of Lowe's invention), several forms of combined heating, lighting and ventilating devices, etc."

Lowe's intention with the showrooms appears obvious; he wanted to show his audience and visitors his blue water gas technology on how it works and how the blue water gas could be used for a variety of purposes. Of importance, of course, the experimental works would provide him with information to understand the chemistry taking place inside the generator and the rest of the process system to guide him to improve, design and operate future projects. What is unfortunate is that little information about his findings available. The show rooms would certainly help him do the sales work, a purposeful and thoughtful setup. Lowe certainly knew the importance of his customers and ways to attract them, probably stemming from his experience with his show business of ballooning. It appears that the experimental works was also equipped with a lecture room and a laboratory

(Auge, 1879). While developing his technology Lowe used the lecture room as a platform to entertain his audience and visitors, from domestic and abroad, by introducing his technology, related improvements and their applications.

Lowe's early commercial activity certainly benefitted from the cheap oil such as gasoline and naphtha, which are typically highly volatile and easy to be fixed into a blue water gas. Although the blue water gas could be used for illuminating purposes a specially designed incandescent lighting device like the one invented by Lowe had to be used, which would be an additional cost to upgrade lighting fixtures to existing customers in the traditional coal gas network. Moreover, a network to distribute water gas to any distance would be another challenge that would take resources to develop. Fortunately enough, the cheaply available light oil and naphtha helps Lowe break the ice. Before gasoline cars were invented light oil and naphtha were cheap and of little value. Carburetted water gas could, therefore, be made by simply passing the blue water gas at end of the process through a pool of lighter cut of oil such as gasoline, and the gasoline vapor would mingle with the blue water gas to become "fixed" for illuminating uses, which could be blended with conventional coal gas distribution networks. Quite a few plants were built to manufacture such an illuminating gas in Pennsylvania and neighboring states. Since Daimler and Maybach invented their first gasoline car in 1890, however, oil market started to shift and the light oil became expensive, further aggravated by the capped production of oil in the region from 1880s. Lowe had to find new ways to accommodate such a shift in the oil market by utilizing naphtha and then heavy oil in order to contain the cost of making the carburetted water gas.

To accommodate such a move, Lowe made necessary changes to the previous design and process by subjecting the heavier oil to more severe conditions of a high temperature and a long residence time in order to crack down the heavy oil enough to small molecules, which would easily be "fixed" in the blue water gas. By 1890, Lowe's CWG process had evolved into a design with a generator followed with double superheaters, one for oil cracking, the carbureting, and the other for superheating the carburetted gas. The Run cycle operation evolved into an up-down-up cycle, which makes the gas making more effective. By this time, Lowe's CWG process had emerged matured as one of the most commonly adopted in the market place, featuring a generator, a carburetor, and a superheater, followed by two waste heat boilers, one for preheating blast air and the other for generating required process steam (Fig. 7.3) (Lowe System, 1915).

Adoption of Water Gas Process

Although Lowe's patents claim that the blue water gas from his invention could be used for both heating and illuminating applications the reality is that he has to resort to oil cracking so to make the carburetted water gas that he or his customers could blend the gas easily into the existing coal gas distribution network, which otherwise the effort to

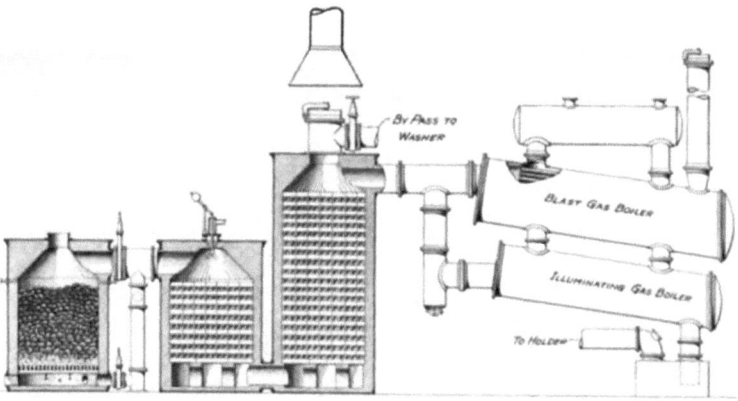

Fig. 7.3 Modern Lowe CWG process

break into the market would be astronomical. From 1874, Lowe started to invest and build a carburetted water gas works to sell illuminating gas such as works in Phoenixville, Conshohocken, and Columbia in Pennsylvania. It worked well and ensured with a rapid growth. By 1879, more than thirty cities and towns had adopted Lowe's CWG process to manufacture an illuminating gas for lighting up about a million people. The cities and towns extended to Lancaster, Harrisburg, Scranton of Pennsylvania, Baltimore of Maryland, Indianapolis of Indiana, Rochester, Utica, Cohoes and Fort Plain of New York, Kingston and Toronto of Canada etc. (Auge, 1879). By 1886, the carburetted water gas had been manufactured from about 144 gasworks located in 32 states in US, in Canada and Cuba (Fig. 7.4). By states, Pennsylvania had 33 followed by New York State with 25, 10 of New Jersey, 7 of Illinois and 6 of Massachusetts. By technology, Lowe's CWG process was deployed in 96 works followed by the Granger process with 26, a technology derived from the former. The balance used a few other technologies. Clearly, most of the gas works were concentrated on the coastal northeastern states, reflecting the well progressed industrial activities in the region, which was benefited from its cheaply available oils.

Lowe's Technology Business

To effectively meet the growing demand, Lowe appears to have adopted a licensing business model early on, under which a licensee agrees to pay a fee to the licensor, the technology provider, in exchange for the rights to use the licensor's proprietary technology to achieve business objective. There are several ways to structure the fee, which can be an upfront lump sum payment, a royalty payment based on the actual operation, or a combination of both. In return, the licensor provides a guarantee of the performance of the technology under the conditions agreed upon between the licensor and the licensee. In

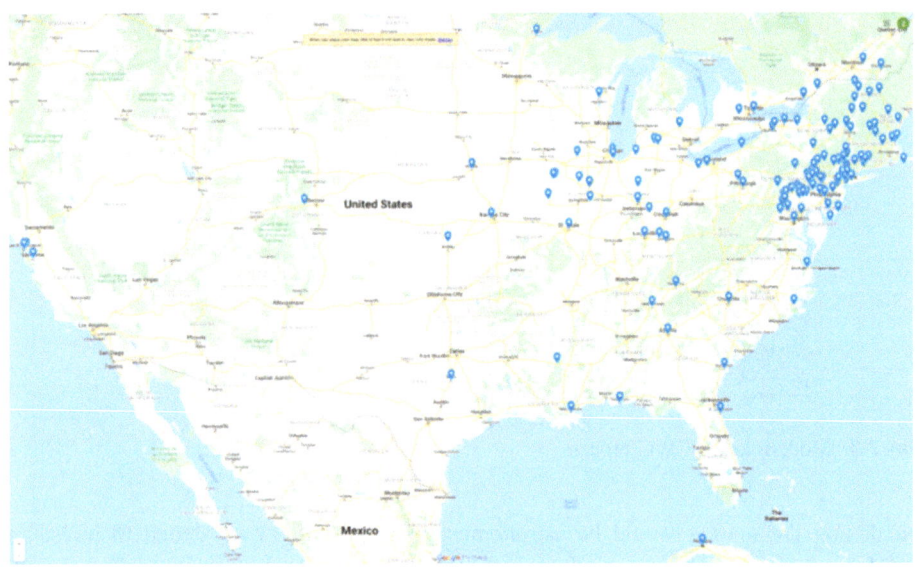

Fig. 7.4 CWG plants in operation in US, Canada and Cuba in 1886 (Data source: Institute, 1886)

addition, the licensor may provide other goods or services in association with the license. It appears that Lowe charged his licensees a royalty payment based on a unit price for every 1,000 ft^3 of water gas, BWG or CWG produced. Such a royalty payment shall be in effect for a specified period of time. In addition, Lowe also marketed his BWG and CWG processes through agencies, for example, Messrs. S. Stevens & Co. of Philadelphia, Granger & Co. of New York, and Messrs. Pierson Bros. etc., to develop and build the Lowe water gas processes in their respective markets. Through these agents, more Lowe water gas plants had been built across the States. Through terms of grant back arranged in the licenses, Lowe had benefitted from improvements either made in design by the agents or by the many operating plants to perfect his technology. The business model seems to work for Lowe well because coal gas making had become sophisticated, and to develop and build a water gas facility, more often than not, additional expertise, domain knowledge, and capability are required beyond water gas making itself. Those agencies brought in their perspective expertise in equipment fabrication, materials selection, downstream water gas treatment, process, utility requirement, and balance of plant, which are all important to the success of any project. The agencies also know their respective market and available resources very well, contributing to the sales of the Lowe's water gas process. In doing so Lowe as the licensor could focus on the technology or process for coal gas making to keep his technology sharp and fresh, reflecting the changing market conditions. Soon, the Lowe water process, BWG and CWG, had become a dominant technology in the coal gas market.

In spite of all the successes, Lowe's business model, however, has a downside. Some of the agents, upon acquiring more experience, also took up patents on their design changes and improvements made to the Lowe process and eventually formed their own brand. The Granger process patented by the Granger & Co. was one of the examples. For a small company like the Lowe's, this would not be an easy job to manage and protect its technology. In 1884, Lowe sold his patent rights of BWG and CWG processes and related business to United Improvement Co. ("UGI") and then moved his family to Pasadena of California in 1887. While continuing his interest in coal gas making and technology development Lowe started to pursue other interests in banking, scenic railroad hotel and an observatory to boost the local economy. The railroad was dubbed the Mount Lowe Railway, recorded by the National Register of Historical Places in January 1993. This last legacy, instead of making any money for Lowe, exhausted pretty much all his money and left him in poverty till his death at the age of 80.

Established in 1882 and headquartered in Philadelphia, UGI is a holding company that owns and operates natural gas and electric utilities. It is the first public gas utility company in the United States, which involved in coal gas making, equipment leasing, and distribution business. UGI also involved in manufacture, sale, and installation of equipment related to the Lowe water gas process (UGI Corp History). A year after acquiring the Lowe water gas technology UGI moved on to consolidate water gas related processes offered in the market at the time including the Granger process. Not long after, UGI had become the dominant player in the rapidly growing coal gas market in America by phasing out the existing coal gas making retorts while building additional capacity to meet the growing demands for coal gas. In 1897, UGI was contracted by the City of Philadelphia to operate and manage the Philadelphia Gas Works, the largest in the States under the City's control since 1841. Aided by the experience, UGI continued to improve the water gas process around equipment manufacturing, installation, and operation etc. The widely used design as shown earlier in Fig. 7.3 and the updated operating Run cycle procedure of up-down-up run for more water gas production and an improved three-layer refractory design are few examples. It is worth noting that the three-layer refractory design inside a gas generator that contains a wear resistant inner layer, an insulating and supporting middle layer, and an expansion outer layer has basically become an industrial standards, which have ever since been practiced within many today's entrained gasifiers. During 1919 right after WWI water gas process provided 60% of the total manufactured gas in the US market covering coal gas, carburetted water gas, coke oven gas and oil gas in the distribution system for a variety of uses such as illuminating, heating and cooking. At the end of WWII water gas process reached another peak time in operation due to energy shortage. By about 1950 water gas almost put an end completely to the retort operation to make coal gas in US and continued its existence well into the 1960s (Committee, 1945) before giving way to the then fast growing natural gas market. As an example, the coal gas manufactured at the Philadelphia Gas Works, supplying 90% of the coal gas distributed to a half million customers within the city limits, had primarily been carburetted water gas by 1940s.

Toronto Consumers' Gas Works

Lowe's water process made its debut at Toronto Consumers' Gas Works in February 1879 where each of the two Lowe's water gas generators produced 75,000 ft^3 of CWG daily. The Toronto Consumers' Gas Works was incorporated in 1848 under a charter granted from the authority of Toronto. To countermeasure the competition from the emerging electricity lighting, the Toronto Consumers' Gas Works, incorporated in 1848, had conducted an extensive due diligence during 1877/88 on Lowe's water gas process. After witnessing the operation of the Lowe's water gas process adopted at quite a few coal gasworks in Canada and the US, the company concluded that it was a sound technology that would benefit its operation. The company then signed a contract with Lowe Manufacturing Co. to purchase the CWG technology right to manufacture CWG to improve the performance at its existing coal gasworks. In its early phase, two sets of the Lowe's generator system were installed with coke from the retort operation as feedstock. Such a success followed with more installation in following years.

Water Gas Beyond America

Before the end of the 1889, the carburetted water gas had been an American phenomenon where the market had embraced the carburetted water gas well by recognizing its superior quality and cost benefits. While America was upgrading its traditional coal gas capacity with more carburetted water gas the other side of the Atlantic Ocean seems not in a hurry to adopt the Lowe water gas process until about 1889. The Chartered Gas Light and Coke Co. (GLCC) in 1889 acquired a license from UGI for two water gas units manufacturing 140,000 ft^3/d of water gas daily at its Beckton gas works, the largest gasworks at the time in Europe. There seem many possible reasons justifying such a delay, but the obvious one may well be the limited oil supply and high prices, which would impact negatively the cost of the carburetted water gas, understanding that most of the oil imports came from America. At the annual meeting of the Institution of Civil Engineers held in 1887 Corbet Woodall, a well-recognized gas engineer and future governor of the GLCC between 1906 and 1916, made a statement about the benefits of the Lowe water gas process.

> Among the incidental advantages (of the Lowe water gas process over the coal gas making retort) attaching to the manufacture of the gas in question were the small cost of plant for a given output, the lightness of the labour, and the reduction of the number of men employed, the fact that the plant was available at its full power within the three hours of lighting up, and, lastly, the better price obtained for coke.

Obviously, the Lowe's water gas process shares the benefits of the Siemens brothers' Gas Producer, which requires light labor, occupies a small floor plan and is capable of churning out of a large quantity of water gas to help peak demand of coal gas. Utilizing the largely available coke from retort operation is another synergy when Lowe's water gas process is applied to a coal gasworks. Of significance from engineering perspective, the much more effective heat transfer within the gas generator than the traditional retort makes the gas generator quicker to start.

To assist the startup of the two water gas units at the Beckton gas works UGI dispatched its chief engineer, Arthur Glasgow, onsite to supervise the ongoing activity. The two units entered into commercial operation in 1890. Soon after, the Lowe's water gas had rapidly deployed in many other local coal gasworks such as the Winsor gasworks and the Tipton gasworks etc., primarily being blended to meet peak demand. By 1897, water gas had reached 50 millions of ft^3, about 8% of total coal gas capacity in Britain (Woodall, 1897).

As a public utility company, UGI's market focus might have been restricted to customers within its territory of America. To capture overseas opportunity Arthur Graham Glasgow (1865–1955), a graduate from the Stevens Institute of Technology in 1885 and his colleague Alexander C. Humphreys (1851–1927), also a graduate from the same institute in 1881, established a partnership in 1892, the Humphreys and Glasgow ("H&G") with offices both in London and New York City, to design, manufacture, and install water gas plants outside America market under a sublicense with UGI. With early water gas deals sealed in Copenhagen, Sweden and Belfast, Northern Ireland, their overseas business took off immediately. By 1904, H&G had contributed 28% of the total sales of water gas capacity to UGI.

At the beginning of WWI in 1913, H&G delivered more than 480 plants to UGI's commercial fleet. Those plants were located in almost every corner of Europe, Far East and Asia Pacific. Between 1903 and 1921, the America was by far the largest market for this native water gas technology, 2.7 times on average than that outside the America market (Fig. 7.5). In 1914, H&G claimed that it along with UGI collectively built 85% of the total water gas capacity worldwide.

H&G made its presence in Shanghai in the early 1900s. The company's pamphlet in 1909 indicated three carburetted water gas plants sold in Shanghai with a total coal gas rate of 2.1 mm ft^3 daily, most likely owned by Shanghai Gas Co. who was the only coal gas operator before the Japanese military built its by-product coke oven plant in Wusong (吴淞区) area in 1938, a few miles north of Shanghai Gas Co.'s gas works. The records of the three carburetted water gas units as claimed by H&G were nowhere to be found locally in Shanghai. Locally available Information on this technology upgrading and others prior to 1930, however, is very sketchy and inconsistent at best. Shanghai Gas Co. most likely started the water gas technology upgrading around 1900. According to Shanghai City Chronicles and several other sources, the company first upgraded two of the eleven sets of five retort benches with three sets of the integrated regenerative

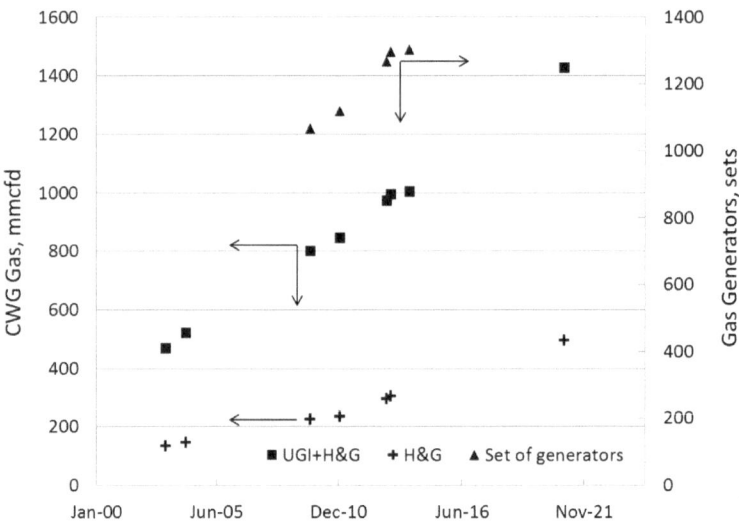

Fig. 7.5 Worldwide Lowe's Water Gas Deployment from 1903 to 1921

horizontal five retort bench, boosting the total coal gas to 19,800 m³ daily, and further the rest by 1904. Then, a new storage gas tank was built in Xi Zang Zhong Lu (西藏 中路) in 1905 followed by another one the next year, totaling 58,000 m³ (2.1 mmcf) of additional storage space. Most likely, it is around this period of time the three water gas units along with purification equipment were built on the original gas works site because the three units would produce a total 2.1 mmcfd of carburetted water gas matching the volume of the two new storage tanks. Water gas and carburetted water gas were important source of coal gas later during the 1930s to improve the quality and cost of coal gas while meeting the growing demand. Such significance might have lost in translation. The Chinese translation of the carburetted water gas, 增热水煤气, when being blended into coal gas, could probably be misconstrued as "a gas to enhance the heating value of the coal gas". Somehow, the value of the carburetted water gas as an illuminating coal gas of high quality, cost effective and easily made in large quantity might have been buried in"增 热水煤气" as something but coal gas, which could explain the lack of local information about the blue water or carburetted water gas process. Of course, records and documents could be lost in the subsequent wars. The key here is that blue water gas and carburetted water gas had become an important part of the coal gas to meet the fast growing coal gas industry in Shanghai and later in China as a whole.

Shanghai Gas Co.'s coal gas making process continued to evolve along with the growing coal gas market. In 1910, the company started to promote coal gas cooking for homes, thirty years after starting the early trials, and the coal gas price was set at 0.14 Silver Dollars/m³. In 1912, coal gas had expanded into shops for metal smelter, rubber tire making and into homes for water heating, increasing at a rapid rate of 11% a year. It appears that Shanghai Gas Co. used imported coals in its coal gas making. When WWI broke out coal prices jumped by three and half times from 5.5 Silver Dollars a ton, significantly increasing the cost of coal gas, the company had to hike the price of coal gas twice in order to stay in business. By 1920, Shanghai introduced a vertical retort, which eventually replaced all the horizontal retorts at the original site. It is during this time that Shanghai Gas Co. mastered the a variety of coal gas making technologies and firmed up a better way to make coal gas, which served as a blueprint for the future Yangshupu Gas Work developed between 1932 and 1934.

Integrated Coal Gas Utility

The end of the American Civil War in 1865 and the German Unification in 1871 started another long period of economic and social development worldwide. Population growth and active industrial activities had pushed cities and towns to expand further, which in turn requires more coal gas not only for lighting but also for heating, cooking, and other emerging industrial uses. By 1890 outside the USA, coal gas making technology via retort remains essentially the same as Murdoch invented a century ago and many of them were still in operation, especially in Europe. Although the traditional retorts had been upgraded with gas firing by leveraging the Integrated Regenerative System, which had significantly cut fuel consumption, improving the overall efficiency, it did not change the fundamentals of how retort operates. No matter whether a horizontal retort or inclined one, it is still a labor intensive operation that requires a periodic loading of coal and off-loading of the resultant coke. Adding more retort benches seems the only option to expand the capacity of coal gas making, which requires space. The problem is that many existing gasworks had no space available because these gasworks originally built on the skirts of the cities half a century ago had now become the center of it. This forced gas companies to relocate its gasworks or to develop new gasworks of more capacity for coal gas such as the GLCC expanded its Peter Street gasworks to Beckton in 1868, a much large space about 11 miles east of the Peter Street gasworks sitting next to the Thames; The Philadelphia Gas Works had to expand its operation at the Market Street to Point Breeze site; and Shanghai Gas Co. had to relocate its gasworks to a much larger site at Yangshupu in 1932, and many more. After a few rounds of expansions then, any further changes, not to mention the expansion, would seem hard to maneuver. Coal gas making industry was definitely in need of some fundamental changes. Although many engineers and chemists tried many ways to address the changes such a dilemma, nonetheless, remained until the beginning of the nineteenth century when a vertical retort was invented.

The vertical retort, continuous or intermittent, is a technology developed in the early 1900 to address the challenges facing the traditional retort operation such as labor intensity, spread out and poor gas quality etc. In a vertical retort operation, coal is fed from the top of a retort via a bucket conveyer and coke comes out at the bottom so that gravity itself does the work instead of labor work as did in the traditional retort operation, horizontal or inclined. Like the integrated regenerative horizontal retort system, a vertical retort system also uses producer gas to fire up indirectly the whole area of the retort chamber where coal, in its downward move, is heated up and gets distilled or carbonized to give off coal gas, which leaves at the top shoulder of the retort. The coal gas thus collected would have a consistent composition because all coal particles are subject to the same temperature profile in moving downward inside the retort chamber. A typical composition of coal gas manufactured from a continuous vertical retort operation is given in Table 4.1 II. A downside imbedded in such a design, on the other hand, is the fact that it does not change the feature of external heating of every retort. Like any other retort operation, therefore, it takes time for the system to work up and down between runs, less flexible in coal gas making. The Woodall-Duckham design and the Glover-West design are two representatives, of vertical retort technologies that had achieved a wide adoption since it was introduced into market in 1903 and eventually phased out the horizontal retort system around WWI. In Shanghai, the vertical retort also found its best use.

The Yangshupu Gas Works was commissioned on February 8, 1934, with its capacity increased to 113,000 m^3 of coal gas daily. In 1931, the Shanghai Gas Co. purchased a piece of land at Yangshupu sitting right next to the Huangpu River. Comparing with the original works site of 9,876 m^2 at 西藏中路 the Yangshupu site was much larger, 22,000 m^2, providing the company an ample space to expand. The new plan called for an integrated solution where the vertical retort would work with both the water gas generator and the gas producer to maximize synergies for the town gas making (Fig. 7.6). Woodall-Duckham Co. of London provided one set of its vertical retort equipment with thirty retort chambers and six sets of its gas producers. The producers used a portion of the coke discharged from the vertical retort to manufacture fuel gas, which was fired to heat up the vertical retort. H&G provided two identical sets of its carburetted water gas generator system, which also used the coke discharged from the vertical retort operation to manufacture either blue water gas or carburetted water gas, depending on actual real time needs. Doing so would maximize the internal consumption of coke out of the vertical retort operation while producing the required fuel gas and water gas. The water gas generator system could manufacture up to 54,000 m^3 of blue water gas of 300 BTU per ft^3 during normal operation or carburetted water gas of 400 BTU per ft^3 during winter time or any time for blending with that coming from the vertical retort operation. The Woodall-Duckham vertical retort system consumes between 150 and 250 tons of coal daily and provides a coal gas between 54,000 and 74,000 m^3 of coal gas for blending to make town gas at a rate of 113,000 m^3 daily for sale. The coal gas from the vertical retort operation has a higher heating value of 450 BTU per ft^3 or higher, depending on its operating conditions. The town gas has a specification of a heating value between 360 and 426 BTU per ft^3, which was typically lower during winter or peak time and higher during non-peak time. Such a spread of heating values provided Shanghai Gas Co.

the necessary wiggle room to maintain the flow of 113,000 m³ daily of town gas while meeting the required specific heating value at different times by conveniently adjusting the blending ratio between the coal gas from the vertical retort operation and the water gas or carburetted water gas. In so doing, the company a few guidelines of operation, that is to maximize the use of the blue water gas as much as possible and whenever it was possible because the blue water gas is cheaper to make and its rate could be easily adjusted by turning down or up of the operation. In addition, the blending of carburetted water gas should be well managed due to the fact that oil, light or heavy, was expensive and not easy to come by in Shanghai. These guidelines had been applied to the design, engineering and operation of the Yangshupu Gas Works.

Once the Yangshupu Gas Works was up running steadily the original gas works was shut down a month later. The 113,000 m³ of town gas was routed each operating day to the existing gas storage tanks. By November 1935, Shanghai had completely replaced its street gas lighting by electric lighting its town gas business, however, continued to grow, and its pipeline network had reached 210 miles.

The Yangshupu Gas Works discontinued its operation when the Shanghai Battle began on August 13, 1937, a month after Japan invaded China. When the Pacific War broke out, the Japanese occupation army immediately took control of the Works and related assets on December 9, 1941. Shanghai Gas Co. did not get it back until September 19, 1945.

In hindsight, it is fair to say that the success that Shanghai Gas Co. had achieved with the Yangshupu Gas Works would be hardly achievable without applying the significant amount of hands on experience, knowledge, and skills about coal gas making acquired over the past few decades. Such an integrated approach exemplifies another excellent best practice in process engineering. The integrated solution clearly made a leap in the coal

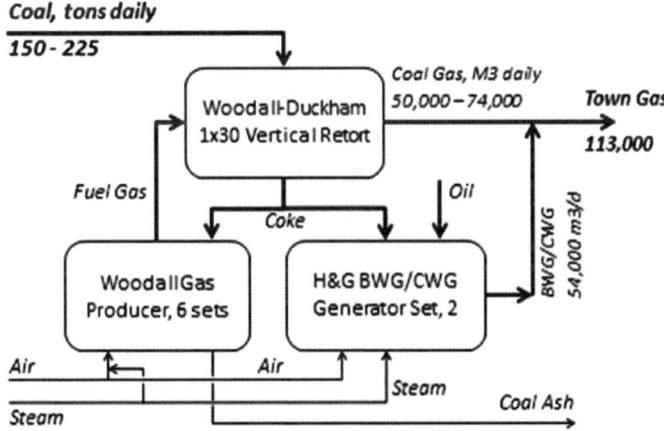

Fig. 7.6 Yangshupu coal gas works gas making scheme (data estimated from sources (Gas Making Technology) (张应莹, 1960))

gas making by pulling in all best available technologies of the time, completely moving away from the 140 year old batch operated, small scale, and labor intensive traditional retort system. The solution had addressed well the needs of the local market by making full use of the installed capacity of the advanced vertical retort system to manufacture high heating value coal gas, which is supplemented as much as possible with the cost effective blue water gas, only carburetted when necessary. Technically speaking, what the Shanghai Gas Co. did at the Yangshupu Gas Works may not be the first one but surely be one of the early unique, creative, and comprehensive engineering piece of work that uniquely fits the local market of Shanghai. In addition, it confirms a rule of thumb or an engineering principle in modern practice that a comprehensive solution is usually necessary, if not required, whenever it comes to coal utilization. Taking synergies by recycling the discharged coke from the vertical retort as feedstock for both the gas producers and the water gas generators to manufacture fuel gas and water gas, respectively, made the whole operation effective and efficient, which would reduce the overall carbon footprint. It is a job well done, not only as the best available solution back then but also inspirational moving forward.

References

America, T. A. (1907). *Navigating the air*. Doubleday, Page and Company.

Auge, M. (1879). *Biographical notes of prominent living citizens*. By the Author

Committee, N. R. (1945). *Chemistry of coal utilization*. Wiley.

Evans, C. M. (2002). *War of the aeronauts—A history of ballooning in the civil war*. Stackpole Books.

Haydon, F. S. (2000). *Military ballooning during the early civil war*. The John Hopkins University Press.

Institute, S. C. (1886). *Report on water gas*. Franklin Institute.

Lowe, T. S. (1867). Patent No. #63404. USA.

Lowe, T. (1911). Observation balloons in the battle of fair oaks. *The American Review of Reviews*, 186–190.

Lowe System. (1915). *Proceedings of the International Gas Congress* (p. 26). International Gas Congress.

Woodall, C. (1897). Carburetted water-gas. In *Proceedings of the Institution of Civil Engineers* (Vol. 130, p. 210). Institution of the Civil Engineers.

张应莹. (1960). *煤气机与煤气炉问答*. 上海: 上海科技出版社.

Gas Engine and Gas Producer

8

Although most of the nineteenth century witnessed the burgeoning coal gas industry, the understanding of what's inside the retort and producer, the chemistry, had made no significant progress but rather remained stagnant. Dalton's atomic theory by organizing a structured scheme that historians tend to think the start of the theory of chemistry did create some excitement early on, chemists and engineers had, however, found it increasingly confusing when it comes to the reconciliation of the concepts of elements, atoms and compounds due to the lack of substances and the emptiness of the atomic theory. The following paragraph is one of the attempts made by Clegg Jr. in his 1841 treatise to explain the chemistry of coal gas making.

> The physical state of bodies depends on two antagonist forces, Cohesion and Repulsion. Cohesion acts on the atoms of bodies, tending to bring them into contact. In solids, then, this force preponderates over repulsion. In elastic fluids the contrary is true; since we find such repulsion among their particles, that were it not for the containing vessels, these particles would be indefinitely separated. In non-elastic fluids, or liquids, the forces are about balanced. It is generally admitted that heat, or, more properly speaking, the cause of heat, caloric, is the cause of repulsion. If we consider caloric as a subtle fluid, we may imagine it to produce the effects of repulsion, by insinuating itself between the atoms or molecules of bodies, and thus counteracting the effects of cohesion.

Obviously, such an explanation is quite mechanical and the caloric as the cause to balance forces between cohesion and repulsion is still as clear as mud. Chemistry was clearly falling behind, if not far behind and still remained in the shadow of Newton's mechanical chemistry. It is interesting to notice that Clegg Jr. used the word 'molecules'; he might have been aware of the debate over the molecular theory at the time, which had caused

Q. Zhuang, *From Coal to Hydrogen*, Synthesis Lectures on Chemical Engineering and Biochemical Engineering, https://doi.org/10.1007/978-3-031-55586-2_8

some tractions, at least, from the perspective of coal gas making. Driven by opportunities presented by the market demands engineers and inventors plunged ahead anyway to continue to improve, modify, and upgrade the coal gas making processes, equipment, and parts etc. While coal gas making continued its trend of growth for lighting up more houses and streets a producer gas emerged as an industrial gas fuel, which further expanded the presence of the coal gas. As a result, making coal gas had become a critical part of the ongoing industrial revolution.

Realizing the advantages and benefits that coal gas could provide, attention had also turned to look into the characteristics of coal gas, physical or physicochemical, to explore additional values and applications. Such exploration eventually led to many inventions. One of them is the gas engine, which soon made the coal gas and then producer gas an indispensable fuel and exclusive in some cases as an economical alternative to the long standing steam engines, especially to small shops and works in needs for small duty prime drivers. In this way, the invention and popularization of the gas engines injected a new life into the industrial revolution, making it more thorough and far-reaching.

External Versus Internal Combustion

Speaking of internal combustion engines or gas engines, we naturally link it to the automobiles powered by gas engines as a prime driver to move around freely. Before that actually happened though, the engines had actually been used to drive machines for a wide range of applications from doing mechanical works to driving dynamos for electricity generation. What is little known is that it is coal gas that made the internal combustion engine possible back in the 1850s. The dream of developing a free standing vehicle, however, can be traced back, at least, to the eighteenth century.

Late in the 1760s, the French military engineer and inventor Nicolas-Joseph Cugnot (1725–1804) was tasked to develop a self-propelled *fardier* that would be capable of moving heavy cannons or equipment for French army. A *fardier is* a two wheeled horse drawn wagon to move military equipment. In 1769, Cugnot built a prototype by adding a specially designed front wheel to the wagon and then placed a Newcomen steam engine on the front wheel, which has a gear mechanism driven by the engine piston. There comes a horseless and self-propelled *fardier*. It worked except that it moved too slowly due to its bulky and heavy nature. Plus, the frequent refueling of coal or wood to keep the boiler up and the water refill would make the likelihood of a working *fardier* hardly practical. The French Army gave up its plan for any further development, but did not abandon the actual prototype of the *fardier* so that it can still be viewed today at the National Conservatory of Arts and Crafts in Paris. According to Cugnot's blueprint, the prototype *fardier* weighs 4,000 kg and has a dimension of $215 \times 725 \times 225$ cm, similar in the dimension of a Ford F150 pickup truck, $220 \times 635 \times 203$, but about twice the weight of the F150. Of course, there is no need to make the comparison of the driving powers.

In England, William Murdock while stationed at Redruth also built a small, three wheeled model automobile propelled by a steam engine mounted on rear wheels in 1784 prior to the time that he laid his hands on coal gas. The steam boiler was heated with a spirit lamp, a type of burner typically used in a laboratory. The model worked. Murdock might have used an improved Watt's steam engine in the model. Murdock continued the development in the next two years but ended without any consequence as both Boulton and Watt did not see any value of the product and discouraged Murdock's efforts for further development. This might have turned Murdock to investigate the coal gas making in his spare time. The interest in developing a self-propelled 'automobile', however, had remained on and off in the market place. In the early 1820s another British inventor Samuel Brown built another steam engine propelled automobile fueled with hydrogen to fire up a steam boiler. By the 1830s steam engine propelled carriage seems to gain an acceptance and become adopted as a way to move passengers around within a city boundary. Those carriages were typically bulky, heavy, slow in motion, and also took time between firing up the boiler and starting the motion, essentially remained unchanged to what Cugnot did sixty years ago even though the steam engines had improved significantly. Steam propelled 'automobiles', nonetheless, survived for more than a century. Some farming tractors propelled with coal fired boilers are still running in some places as a showcase of the good old days.

Coinciding with the time of using coal gas for lighting, the scientific world had become more interested in exploring further on those gases, inflammable or not, and their related properties. There seems a school of thoughts, especially in the European continent, to utilize the explosive nature of an inflammable gas such as hydrogen or coal gas when being subjected to fire or a spark under certain conditions so that the sudden expansion of gases would create power. If the violent expansion could be managed to do work such gaseous fuels would transform external combustion to internal combustion, which would make a new type of engine of much simpler and more efficient than the steam engines because fuel gas could directly be introduced into the steam cylinder and expand when subjected the fuel gas to a spark. The subsequent explosion would force a piston to do work, just like what steam does to the piston. What a wonderful engine it would be by shooting two birds with one stone, eliminating the fireplace and the boiler in a steam engine system. In 1801, Philippe LeBon while showcased his invention *Thermolamp* also contrived a design of a gas engine by using gas fuel. The design proposed to use a spark to ignite the gas fuel, probably based his experience with the lighting experience with his *Thermolamp*. But LeBon never actually build the engine. Based on a similar design to LeBon's Isaac de Rivaz, a Frenco-Swiss engineer, built a prototype engine powered by hydrogen ignited with a spark in 1804. He later in 1807 applied this prototype to a carriage with hydrogen administered manually via a pipe connected to a balloon where hydrogen gas was stored. Rivaz was granted with a patent both in France and Swiss. These efforts, however, seem stopped short in further development.

On the scientific front, in the meanwhile, some chemists and physicists in France started to look into those gases and their expansion behaviors. One of them is the French Scientist Joseph-Louis Gay-Lussac (1778–1850) who experimented extensively with gases and observed their volume changes at different temperatures and pressures in his early career. He concluded that all gases expand equally when subjected to the same conditions, temperature and pressure, which contributed to the so called Charles-Gay-Lussac Law. Gay-Lussac's other works also had a significant contribution to the debate on the molecular theory that lasted for decades. During his teaching at the Ecole Polytechnique, his chemistry class influenced many curious minds like Sadi Carnot who became interested in gas chemistry and the steam engine. To a certain extent scientifically, the early part of the nineteenth century seems to have paved a solid fundamental ground for a better engine to be created; example are the available Alessandro Voltaic battery, the Carnot Cycle about the principle of an engine design and operation, Joule's clarification on caloric, and the subsequent establishment of the thermodynamic principles by Lord Kelvin and Clausius, which marks the start of the physical chemistry. But nothing had happened until 1859 when Lenoir integrated all the pots and pans into an existing double acting steam engine, introduced coal gas into it, and ignited the gas with a spark an internal combustion engine came to exist. By that time in Paris, *À une époque où l'on installe dans les immeubles le gaz à tous les étages* (at a time when buildings are installed with coal gas on all floors) (National Conservatory of Arts and Crafts, Paris). The readily available coal gas naturally became the convenient fuel that fires up the gas engines.

Coal Gas and Gas Engine

Jean Joseph Étienne Lenoir (1822–1900) is a self-made Belgium inventor. He came to Paris at an age of 16 and soaked himself in what Paris could offer as the center of chemistry and science. Lenoir obtained his first patent in 1847 while working at an enamel shop, using an oxidative chemical process to make white enamel. He sold the patent to his boss, the shop owner, who gladly paid for it. Eventually, Lenoir made his living by selling his inventions such as an electrolytic plating process, a mechanical kneading machine, an electric signal and a brake for railroad, a controller for dynamos, and a water meter etc. His ingenuity and hardworking reimbursed him well. When observing Cugnot's *fardier*, showcased at the École Centrale Paris, Lenoir was becoming fascinated and believed that he could make it a better one. It seems that Lenoir had exposed to the school of thoughts on the internal combustion engine and its earlier trials while taking free lessons offered by the École Centrale. He realized that the steam engine could be made into a better one without using steam. Then Lenoir decided to use his own money to develop a gas engine by using the readily available coal gas in Paris. He must have realized that the powerful expansion of coal gas upon a spark in a confined space could be directed to do certain works. At his friend's workshop located in the center of Paris, Lenoir converted a double acting steam engine cylinder by introducing the coal gas at a slight pressure into each side of the cylinder alternately and ignited the coal gas with a spark plug of his invention. The cylinder piston was hooked directly to a crankshaft driving a flywheel and the coal gas was administered with a slide valve controlled by the crankshaft via an eccentric. The ignitor plug was powered by a battery. The first gas engine was born, a two stroke engine running smoothly. In late 1859, Lenoir filed a patent for the gas engine (Tietz, 2021). Such unique engines attracted a significant market interest. After breaking up the ice in May 1860, Lenoir sold 138 gas engines of different sizes up to 4 HP in the next month. What's interesting is the fact that those engines were not mounted to automobiles, but instead, used in shops providing prime drivers to small machines like printing presses, lathes, and water pumps. Although Lenoir had tried his engines on a boat and also on a three wheeled automobile in the following years the attempt did not go far due to the lack of an appropriate fuels supply at the time. Obviously, it is simply impractical to carry enough coal gas on a moving object, and those engines might have been too small for the automobile as well.

In hindsight, the easy access to coal gas supply might have played a critical role in the success of Lenoir's creation of the two stroke engine and the engine business. Back then, another available fuel gas was hydrogen, typically produced from reacting iron turnings with dilute sulfuric acid, which the latter was be manufactured with a century old technology, the Lead Chamber Process. Although such a fuel gas was of high purity and well recognized as a fuel and an ideal gas for ballooning, this approach had yet been proven on an industrial scale, setting aside the high cost to produce it. Although a few other technologies were emerging at the time such as electrolysis to split water for hydrogen generation and reacting red hot iron or coal char with steam their viability had to go through additional development. As a matter of fact that the electrolysis for hydrogen production was too expensive and only operated in a small scale back then; it is still an emerging technology today. Practically, it was coal gas, rapidly growing, manufactured at a large scale with a competitive price that enabled the sales of thousands of gas engines. Otherwise, it might have taken a while longer for the internal combustion engines to be deployed commercially.

Two stroke gas engines have many advantages for stationary applications, quick to start up, compact, cost less, and occupying a much smaller floor space compared to a steam engine of similar size. This does not mean that it was perfect. There were a few drawbacks inherited from its design such as low efficiency causing a high consumption of coal gas, ineffective air cooling of the cylinder causing severe wear and tear that requires frequent lubrication, and more. Nonetheless, it seems favored by the small shop owners who did not have enough space, could not afford of a steam engine or did not need a big steam engine. It served the market of small sized gas engines well at the time until a better product would then emerge.

Coal Gas and Four Stroke Engine

Nikolaus August Otto (1832–1891), also a self-made German engineer and inventor, was drawn into Lenoir's gas engine while working as a sales representative in Cologne. Like what Lenoir did to Cugnot's steam engine Otto noticed the drawbacks of Lenoir's engine as well and believed that he could improve it. In 1864 Otto met with Eugen Langen (1833–1895), a German businessman, who took an interest in Otto's idea to improve Lenoir's engines. The two then formed the N. A. Otto and Cie. with Langen's money and Otto's expertise, to develop and manufacture an improved gas engines. The partnership proved successful and in three years they had made their much improved gas engine that won a gold medal at the Paris Exposition 1867. Their engine was much more efficient, consuming less than half of coal gas that Lenoir's engine did. This kicked off their fast expansion. After moving their factory to Deutz of Cologne suburb in 1869 they incorporated their operation as the Gasmotoren-Frabrik Deutz AG in 1872 in anticipation of increasing orders of engine sales. The biggest milestone for Otto did not come until 1876, a time when Otto revolutionized the gas engine design by having a four stroke cycle instead of the current two stroke cycle, following the principle the Carnot Cycle. The four stroke cycle engine or the Otto Cycle engine took another gold medal at the Paris Exposition in 1878. In the following years, tens of thousands of the Otto Cycle engines were sold.

Compared with two stroke engine where each stroke produces power, Otto's engine has four strokes, suction-compression-explosion-exhaust among which only the third stroke produces power. Designing this way Otto uses the flywheel to compress the intake fuel of coal gas and air mixture in the second stroke so that the next stroke, explosion, would do much more work during expansion because such compression significantly increases the potential energy of the intake fuel/air mixture. Before this time, all two stroke gas engines had been operated with an atmospheric mixture of fuel gas and air. For some time, actually, it had been believed that fuel gas compression would help boost power generation. The issue, however, is that no tools or mechanisms were available to achieve gas compression. The slight pressure from the coal gas supply was actually created way upstream from retort operation and was of little value to such an expansion power in an engine environment. By leveraging a flywheel, Otto was able to increase the thermal efficiency of his four stroke engine to 12–15%, relative to 3–5% of the early two stroke engines. Such a leap in technology had certainly boosted their business precipitously. Although Otto lost his patent in 1886 thousands of the Otto Cycle engines were sold.

The creation of gas engines not only increased demand of coal gas but also helped coal gas expand further into industrial sectors where tens of thousands of gas engines had been deployed as the prime driver for a variety of tasks. The Otto engine design has remained one of the most important inventions to this modern day.

At the time Otto and Langen continued their strategy to position their engines for stationary use because of the readily available access to coal gas in cities and towns.

Gottlieb Daimler and Wilhelm Maybach who joined Otto at Deutz back in 1882 decided to leave the company to pursue their interest in building the *fardier*. When they placed an Otto Cycle engine onto a horse drawn four wheel carriage in 1890 there was the first automobile on four wheels. Thereof the automobile industry was born. What made it possible was a liquid waste stream, gasoline that became available at the time from the distillation of oil or coal tars. So, the Otto Cycle engine went on and eventually led to another revolution in establishing the automobile industry, which is still going strong as of today. By 1920 there were about 9 million gasoline fueled automobiles running on the streets. The number had gone to 28 million gasoline powered automobiles ten years later. The emerging automobile industry changed the landscape of oil market, which created an opportunity for coal gas four decades later.

On the other side, among the tens of thousands of Otto engines sold most of them had been fueled with coal gas since the beginning. To expand the market, another type of coal gas, a producer gas would be developed to fuel gas engines in a few years. The producer gas was more cost effective and off-grid supply, made possible by Joseph Emerson Dowson.

Dowson Gas Producer

Joseph Emerson Dowson (1844–1940) was a British civil engineer received his education as a civil engineer at the Lycee De Versailles and Dulwich College. After graduation, Dowson joined his father's engineering firm where he acquired extensive engineering and project experiences through different projects and roles on job. From 1876 as a freelancer, Dowson became acquainted himself with the furnaces for metal heating or ore smelter fired up with producer gas. To better understand how the furnace system works Dowson dived into the system and the chemistry behind it, and backed out of it, filed a patent in 1878. It was a small and compact system of a gas producer targeted to make clean and cheap fuel gas for gas engines where owners had either no access to coal gas supply or had limited available space to install a steam engine. This is about two years after Otto invented his four stroke engine, the Otto Cycle engine. It appears that Dowson envisioned that his product would have to be designed for a captive use dedicated to the niche applications. To be successful, Dowson realized that the fuel gas from his product would need to be free of tars and fine particulates which tend to plug small pipes, cocks and/or burner tips, different from the large scale furnace applications where a raw fuel gas without any treatment could be used like what the Siemens brothers did in the open hearth process and the high quality glass making. What happened next proved his vision in next few years and ever since Dowson had dedicated the rest of his career to the development and application of gas producers. He also is one of the early few who conducted R&D on gas producer, coal, and coal gasification to understand the chemistry and mechanism taking place within the gas producer.

Similar to Siemens brother's Gas Producer, Dowson's gas producer is a positive pressure gasification system to produce fuel gas with only a slight pressure over the atmosphere. What is different is that Dowson's design used anthracite coal to minimize the formation of tars inside the producer so to help maintain the system compact while avoiding other related potential operational problems. He deployed a boiler to generate superheated steam, which carries air at a nozzle into the producer to establish a positive pressure. Siemens brother's Gas Producer achieved it with an air blower. The injection of superheated steam would help boost efficiency of the gas production. Dowson's design is a compact system comprising of a boiler, a producer, a hydraulic box and a gasholder, a total package (Fig. 8.1), essentially a scaled down version of a complete coal gas works. The gasholder is designed to play multiple roles including cooling, cleaning and storing the gas fuel. The level of the gasholder is automatically tied to the producer operation by controlling air and steam inflow. At the end of the day the built final product would occupy as small floor plan as possible. To make this compact producer work, however, Dowson had a very specific requirement on what feedstock that would go into the producer.

When Dowson put his compact producer system into trials in 1879, he requested a local maker of the Otto Cycle engine to have a try with the fuel gas produced from his compact producer system. Fearing the poor quality of the fuel gas, however, the local engine makers originally declined to do so because of the low heating value. Engine

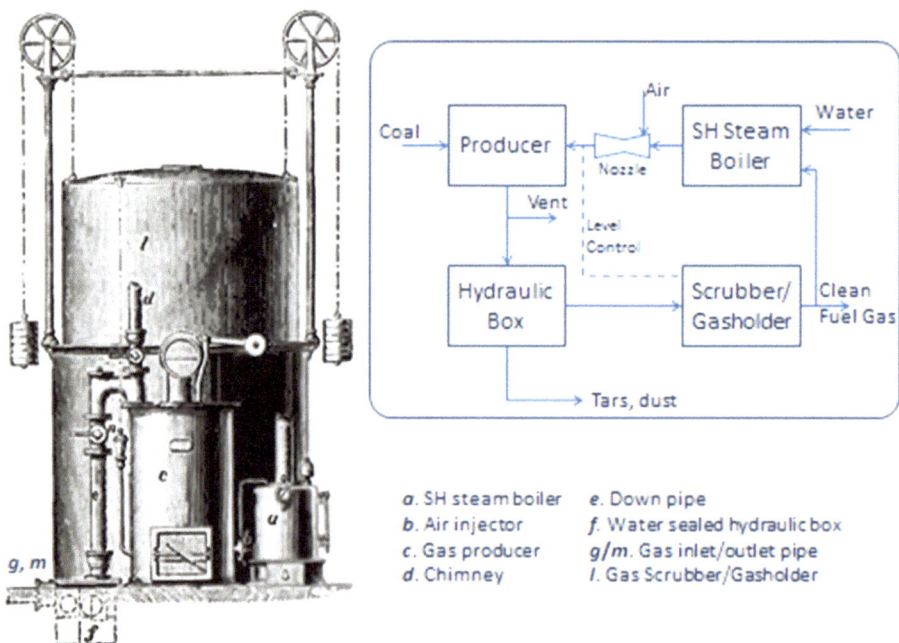

Fig. 8.1 Dowson Producer (1878–1881) and process scheme. *Photo Source* Larter (1920)

makers had all the reasons to be careful because the Otto Cycle engines had been running on the traditional coal gas, which has a heating power of nearly 4 times that of Dowson's producer gas or producer gas in general. In addition, there seems a normal perception by many including chemists at the time that any fuel gas containing a high inert gas such as nitrogen or carbon dioxide would not result in an effective burn in an engine environment. After all, it is a special field of little known. Here is what Dowson stated at the beginning of his book back in 1906, a well put statement telling about Otto and Langen's business from another angle back then, a time when Otto and Langen had just started to sell their early models to the market.

> I am also indebted to some of the makers of gas-engines. Not unnaturally they were at first disposed to think that it would not be to their interest to work their engines with producer gas, as it was so much weaker than town gas (coal gas): the reputation of the engine itself was not then fully established, and they were fearful lest by using producer gas they might add to the difficulties with which they had to contend. That was in the days of small engines, when the consumption of gas was not great, and when the cost of the gas was of minor importance. Later they realized that the future of gas power depended largely on producer gas, and I had the advantage of their co-operation.

In order to make his design work Dowson understood that he would have to use high grade coal such as anthracite or coke as feedstock in order to minimize tar formation. With anthracite in the trials, Dowson producer yields a fuel gas with 49% in volume of nitrogen and high hydrogen of 18.7% (Table 8.1). Even though Dowson's producer gas has a higher heating value of 160 BTU per ft^3, higher than Siemens producer gas of 125 BTU per ft^3, but still significantly lower than a coal gas of 550–650 BTU per ft^3 with which the Otto Cycle engine had been designed. The low heating value of a producer gas is an inherent feature due to the use of air. To Dowson's surprise, however, Crossley Brother, who held a license right for making the Otto Cycle engines in England, made a 3 BHP Otto Cycle engine available to try his producer gas in 1879. The results made all happy; the engine worked just fine. To make sure its reproducibility, more tests ensued into the next year. In 1881, Dowson presented his trial results at the Meeting of the British Association of the Advancement of Sciences in York, and then showcased the compact producer integrated with a 3 BHP Otto Cycle engine at the Smoke Abatement Exhibition the same year. The committee tested the integrated product, and it scored 3.26 BHP at 156 rpm. Sir William Siemens presented Dowson with a gold medal to reward his achievement as *"the best method for the utilisation of fuel as a heating agent for domestic and industrial purposes, combining the utmost economy, with freedom from smoke and noxious vapours"*.

Such an integrated product created a leveled play field for both Dowson and Otto, a win–win situation to penetrate each other's market. In the meanwhile, it also helped the Otto Cycle engines gain additional market space from the steam engine business. Prior to

Table 8.1 Typical compositions of producer gases

	Siemens brothers' producer gas	Dowson gas	Mond gas
Hydrogen	8	18.73	24.8
Carbon Monoxide	23.7	25.07	13.2
Methane	2.2	0.31	2.3
Carbon Dioxide	4.1	6.57	12.9
Oxygen	0.4	0.03	0
Nitrogen	61.5	48.98	46.8
Olefiant		0.31	–
Total	99.9	100	100

Note Data for Siemens using a mixture of caking and non-caking coals; Dowson's with anthracite. Mond gas tested with a 25 HP gas engine at the Winnington site on July 5, 1894

that time, a traditional coal gas fueled gas engine beyond 20 HP would hardly be competitive against a steam engine. The marriage of a gas producer and a gas engine kicked off a remarkable journey of growth for both technologies. In about a half century, Dowson had made the gas producer one of the most widely diversified and adopted gasification processes. By 1889, the producer gas fueled gas engine had reached 60 BHP and further became competitive up to 120 BHP in 1894. Then, aided by another producer technology developed by Mond, the producer gas fueled gas engines, reached larger than 1000 HP before 1910 and 2000 HP (1.5 MW) and higher by the 1920s. Gas engines had become an affordable competitor to steam engines as prime drivers. In addition, gas engines also found more other applications with industrial processes, steam boilers, transportation carriages, farming tractors, onboard river and ocean going vessels, and so on. All these would not be impossible without the producer gas, a clean and convenient gas fuel derived from coals.

The market for gas producers in the gas engine field, however, is not without limit. Large gas engines became bulky, for example, the Cockerill single cylinder Otto Cycle engine rated at 700 HP weighs more than 160 tons including a flywheel, and the bore size of the cylinder reached 51.2 in. (Lawton, 2011). When the gas engines become larger, say more than 1000 HP, its pistons tend to suffer a higher thermal stress, which would cause significant tears and wears. Somewhat, gas engines had become constrained by materials and as well as the technology itself, and subsequently impacted the market of gas producers. In the same time, the emerging diesel engines from 1920 and onwards also restricted the further deployment of gas engines.

To counter the market reality Dowson established his company, the Economic Gas and Power Co. in 1887 in London to continue to improve his gas producer technology while exploring new industrial applications such as for furnace heating, reheating, annealing, and heat treatment etc. for iron and steel works. By 1903, gas producers fueled gas engines had collectively reached about 60,000 HP. Assuming average 120 HP each running out there, there would be 500 gas engines fueled with the producer gas.

In 1910 Dowson merged his business with the Mason's Gas Power Co, a firm established in Manchester in 1905 and specialized in fabricating furnaces of all sort, to form the Dowson and Mason Gas Plant Co Ltd. in Manchester with office in London, which operated until 1960s. Dowson acted both as an engineer and managing director.

Understanding the gas producer or gasification a new field without much prior knowledge and experience to draw upon Dowson also charted the new company with a strategic focus on innovation and R&D activities as a way to tackle issues and challenges of both fundamental and technical that surfaced from ongoing projects. This is in addition to its routine business to make sales of his products and services. This pioneering R&D activity was the earliest fundamental work on coal gasification recorded (Guide G.). Such an R&D initiative seems to have generated a significant amount of fundamental knowledge which Dowson shared in one of his books, *"Producer Gas"*, first appeared in 1906 and had run into at least four editions by 1920. In the book Dowson and his assistant A. T. Larter started with the fundamental of a gas producer theory with content including chemical reactions involved, their equilibrium and impacting factors etc., and then followed with a systematic analysis of the producer system, gas engine and their interactions and improvements for the best combining results. The discussions were substantiated with significant amount of trials or experimental data. The book is still relevant today. Alongside his routine work, Dowson found time to write papers and to present them at conferences to share his progresses and updates of gas producers and gas engines. He also actively engaged with many professional institutions such as the Institution of Civil Engineers and the Institution of Mechanical Engineers. Although retired from his active duty in about 1916, Dowson had been active professionally even at the age of 93.

Producer Gas and Water Pump

Dowson's gas producer as one of the most diversely developed technologies had been adopted in a wide range of fields wherever an economic solution for mechanical drivers is needed. One of the genius inventions is to apply the producer gas to pump water for irrigation in the early 1900s. Herbert Alfred Humphrey (1868–1951), a British engineer, had acquired extensive experience in gas producer and gas engines while working at the Brunner, Mond and Co. He invented the Humphrey pump around 1909 while working as a consulting engineer. Not like any previous pumps driven by steam engines and having many moving parts Humphrey used no moving parts at all by leveraging the explosion force of a producer gas to move water to a higher elevation. The Humphrey Pump system looks like a U shaped piping infrastructure with three components, a gas producer, a gas explosion chamber, and a U shaped large bore pipe (Fig. 8.2). The explosion chamber is equipped with producer gas inlet, exhaust outlet, and a sparking mechanism. The bottom

Fig. 8.2 Humphrey pump
designed in early 1900 (Larter,
1920)

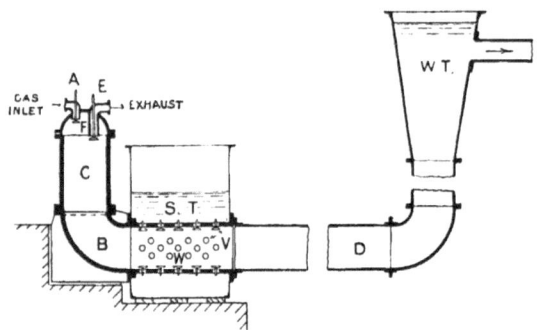

of the U shaped pipe sits in a reservoir, below the surface of water, having holes for water intake and each of the holes are capped with a valve from inside the pipe, called mushroom cap. The high end or the other side of the U pipe is the water exit. It works like a gas engine in principle but uses water itself as a "piston". Upon each explosion the "piston" would push water already inside the U out toward the right side to the water exit and at the end of the explosion the exhaust outlet opens to let out the exhaust while in the meantime water is drawn in via the many mushroom caps to fill the pipe with a necessary water level. Then, the producer gas inlet opens to introduce fuel gas, followed by next explosion to start another cycle. The cycle operation goes on and on, moving water to a higher location. It is a smart and ingenious way to use a gas producer to pump water. In June 1913 five Humphrey Pumps were installed at Chingford, Essex to pump water for irrigation for the community. There were four large pumps and a small one; each of the large pumps were designed to move 40 million gallons of water daily to an elevation of 25–30 ft above and the small one half of the capacity. Four Dowson gas producers were equipped to produce the required fuel gas, three larger ones and one small; each of the three large producers were capable of gasifying 370 lbs. of anthracite daily and the small one at half of the capacity (Larter, 1920). Since then quite a few number of the Humphrey Pumps had been installed in similar applications worldwide such as in Egypt, the United States of America and Australia. The one in the US was installed in Del Rio, Texas in 1914.

The Humphrey Pump proves to be a simple, robust and efficient tool that is quick to start as long as the producer gas is available; it requires minimum maintenance as it has no moving parts that would otherwise require frequent lubrication to minimize the normal wears and tears as an engine does. A downside of it, if any, is that the pump head has to be positioned under the level of the source water because the Humphrey pump has no suction force as a conventional rotatory pump does, which restricts a wider use in some cases as the explosion chamber would be prone to flooding. As long as the case allows, nonetheless, the Humphrey pump would be an attractive solution to which the gas producer is certainly an enabling component.

To recap, the dream to create a *fardier* to drive around freely has been there for a long, but the journey to make the dream a reality took way more than a century. The core at issue appears to be the availability of a practical and convenient onboard energy source. Steam engine of either coal fired or hydrogen fired turns out no fit to the desired *fardier*. While it may not be an overstatement that the coal gas is what made the Lenoir's creation of a gas engine possible it would be unfair to ignore that fact that the scientific knowledge developed before the time about gases and its use for works, the principle of an internal combustion engine and its potential benefits, and the evolving knowledge on heat and the principles of thermodynamics are what equipped Lenoir and Otto with the necessary engineering aptitude to unleash their passion and ambition to realize the first and critical step by building the heart of a *fardier*. Realizing the values of the gas producer available at the time Dowson captured the opportunity of the emerging gas engines. By developing a compact producer system Dowson was able to make a small gas engine a stand-alone operation, no longer in need to access the network or grid of the coal gas, another step forward toward building a practical *fardier*. Although it is Daimler and Maybach who finally delivered the four wheeled fardier in the 1890s the principle of Dowson's compact producer was widely used to drive the *fardier* a few times over the next a half century. Actually, during the two wars and after WWII a large number of small gas producers of suction type had been mounted to automobile vehicles, cars and trucks, in countries where the supply of gasoline was limited. For example, Sweden mounted individual producers to 90% of its vehicle fleet around 1942–43. Most of European countries set up stimulus policies to encourage the use of producer gas for automobiles. So did Japan. In China during the 1950s public buses running on the streets of Beijing carried a giant bag filled with coal gas on top of the buses. In Shanghai, the back of each of the about 160 buses running on streets carried a gas producer fueled with anthracite.

Along with the steam engine technology, the wide range of industrial deployment of producer gas to drive gas engines and for industrial process applications has moved the ongoing industrial revolution to another level, kicking off the electrification era from 1880 and the era of motorization ten years later in the 1890s. The domino effects of these events have completely changed the ways of public lives. In the end, the electrification enabled by Thomas Edison eventually replaced the traditional coal gas for the lighting industry before WWI. The motorization created by the internal combustion engine invented in the 1890s led to the birth of the automobile industry by putting out millions of cars, trucks and buses on roads that are still popular today, which created another opportunity for coal gas before the start of WWI, an era of synthetic chemistry.

Coal gas has more to contribute.

References

Larter, J. E. (1920). *Producer gas.* Longman, Green and Co.

Lawton, B. (2011). A short history of large gas engines. *International Journal for the History of Engineering & Technology*, 79–107.

Tietz, T. (2021, January). *Etienne Lenoir and the internal combustion engine.* Retrieved May 2023, from Daily Blog on Science, Art and Tech in History: http://scihi.org/etienne-lenoir/

Fertilizer and Coal Gas Making

9

Long before the Helmont's controversial experiment in 1640, the use of manures of human and animal origins in crops growing had been practiced way back to the old civilizations such as in Egypt, China, and India etc. Until the time when Justus von Liebig (1803–1873) published his book *"**Die organische Chemie in ihrer Anwendung auf Agricultur und Physiologie (Organic Chemistry in Its Applications to Agriculture and Physiology)**"* in 1840, however, the knowledge about fertilizer was limited and the chemistry behind it simply nonexistent as a shared knowledge in a systematic manner. In his book Liebig pointed out the difference between energy and nutrition to the growing of crops, and went further to explain the role of nitrogen.

> We may furnish a plant with carbon acid, and all the materials which it may require, we may supply it with humus in the most abundant quantity, but it will not attain complete development unless nitrogen is also afforded to it; an herb will be formed, but no grain, even sugar and starch may be produced, but no gluten.

Nitrogen as the most critical component regulates energy flow within crops and plants during their development to form proteins. Liebig's work was significant in a way that it laid the ground for modern agriculture by essentially establishing "the law of minimum" that made Liebig the father of the fertilizer industry, and eventually led to the agricultural revolution. It also aroused significant interests in fertilizer making for agricultural purpose in England and continental Europe.

Before artificial ammonia became available, sources of nitrogen fertilizers were limited. Many small shops were set up to extract nitrogen fertilizers from animal leathers and turfs. Since 1840, European countries and England had imported most of Peru's guano, which had nourished the Inca civilization for centuries, to meet their growing fertilizer

Q. Zhuang, *From Coal to Hydrogen*, Synthesis Lectures on Chemical Engineering and Biochemical Engineering, https://doi.org/10.1007/978-3-031-55586-2_9

demand. Companies like W. R. Grace & Co. established in 1854 in Peru as a trading company in the Guano business during the booming time. 12 million tons of guano had been mined during the following thirty years (The Great Peruvian Guano Bonanza: Rise, Fall, and Legacy). The trade made Peru immediately the dominant supplier of fertilizer to the world before the resources were drained out.

Although engineers and managers at coal gas works had been aware of the formation of ammoniacal liquor from the early days of coal gas making there was no knowledge linking its potential value as a fertilizer. To the contrary, the ammoniacal liquor was a waste stream, which emits a strong pungent and irritating smell, and was simply dumped into nearby rivers or creeks. By 1810 as Accum pointed out in his treatise (Accum, 1819), some works started to use the liquor to manufacture chemicals such as ammonia carbonate or muriate of ammonia (ammonia chloride) primarily for medicinal uses. As time went on, coal gas operations learned more such as that caking coal like the Newcastle coal in northeastern England tends to produce a strong liquor than non-caking coals. For example, a chaldron (5,264 lbs.) of Newcastle coal produces about 180–220 lbs. of ammoniacal liquor. A gallon of the strong liquor could neutralize 15–16 oz of sulfuric acid of 1.84 density while a gallon of weak liquor from a non-caking coal could only neutralize 8–10 oz. Clearly, a caking coal gives off about 50% more ammonia than a non-caking coal. Clegg in 1841 cited a case that Mr. Watson, the manager of the Gas-works at Kirriemuir, Scotland tried to sprinkle the ammoniacal liquor over a piece of grassland in his neighborhood, and found out that the sprinkled grass would grow much faster after a fresh cut than the grassland applied with other manures (Samuel Clegg, 1841). Back to around 1820, GLCC started to look into this waste liquor by building distillation equipment to produce ammonia salts such as ammonia sulfate at its gasworks on the Peter Street and the Brick Lane. Such efforts continued on and off in the following decades and the by-product salts were sold to farmers as cheap fertilizers until the 1870s when GLCC commenced its by-product works at its Beckton gasworks in 1879 (Townsend, 2003). This seems partly due to the ample supply of the Quano imported from Peru and later the saltpeter as fertilizer were preferred by farmers for their quality and convenience to apply to soils.

Entering 1880, fertilizer demands became stronger. With technology processing ammoniacal liquor to ammonia sulfate improved the Beckton Gasworks started to see that its sales of the by-products ammonia sulfate had become an important revenue stream to the overall operation. Many other gasworks followed the suit and soon coal gasworks became the second major source of fertilizer, next to the saltpeter import. Chemically, ammonia sulfate turns out to be a better fertilizer because it is not only convenient to use but also easily absorbed by plants as a source for both nitrogen and sulfur that plant needs to develop proteins. Such a change was certainly made possible by the improved technology to produce sulfuric acid since the 1830s in Britain by using the sulfur recovered at the coal gasworks. By 1890, there were several artificial sources of ammonia sulfate in addition to the saltpeter import. The three major ones are from the coal gasworks, the coking plants for iron and steel making, and the shale cracking among which the coal gasworks was the major contributor. The three sources provided a total of 191,000 tons of ammonia sulfate in England in 1896. The number increased slowly in each of the following years and reached 220,000 tons in 1901 among which coal gasworks contributed 68%. It sold at a good price as well, a ton of ammonia sulfate commanded a price between 10–11 Pounds in 1901 with a trend still creeping up (Co, 1903).

What is going on in the fertilizer market had attracted a curious mind, Ludwig Mond who became interested early on in his career that led him to land on one discovery after another that he would have eventually developed them into several empire businesses.

Ammonia and Soda Making

Ludwig Mond (1839–1909) was a prolific chemist and industrialist whose curiosity and ingenious creativity had driven him on a constant lookout for new things, ideas and discoveries, which could be transformed into business opportunities. During his life, Mond along with his partners and assistants had developed several technologies and processes that crossed a wide field of scientific disciplines and turned several of them into industrial empires of long lasting impacts. To name a few, they are the Solvay process that revolutionized the soda making, the Mond gas process co-producing fertilizer and fuel gas, the discovery of metal carbonyls and its development of a nickel refining process, and the exploitation of the hydrogen fuel cell technology for electricity generation etc. The significance of these developments and discoveries are far reaching and deeply rooted into the fabric of the industrial revolution, economic and social lives around the time entering the twentieth century. To stay focused, however, this section will focus on the Mond gas process, relevant to coal gas making, while in context provides a brief account of others relevant.

In 1855, Mond studied chemistry under Robert Bunsen at the University of Heidelberg. Without obtaining the Doctor's degree Mond left the university and worked at a few different chemical works such as a factory making verdigris near Mainz and a Leblanc soda works at Ringenkuhl near Kassel. During the early nineteenth century sulfuric acid and soda were two important products that chemically manufactured. Soda had become a hotly sought product for soap making, textile industry, glass and paper making in addition to bread baking. During the 1870s Britain manufactured 200,000 tons of soda, more than the aggregate made by the rest of the world.

The Leblanc Process as the only process available to make soda, however, was a wasteful process that only yields 16% of soda by weight out of the feedstock of coal, salt, lime and sulfur source. It is also a heavily polluting process with emissions such as hydrogen chloride vapor and black ash (calcium sulfide), presenting a significant hazardous working environment that is hardly to fathom today. In the Leblanc process, sulfur as an intermediate chemical is necessary to make sulfuric acid, and needs to be recycled, which caught

Mond's interest. Back to Cologne where his uncle lived in 1861 Mond continued his investigation to recover sulfur from the waste stream of calcium sulfide while working at a factory to produce ammonia from spent organic wastes such as leather. In the same year, Mond filed a patent about it in France, and Mond went to England in 1862 to demonstrate his invention at the John Hutchinson & Co., a Leblanc soda factory established in 1847 in Widnes, Lancashire. With a few years of additional improvements, Mond was able to sell his technology to many soda plants by recovering up to 50% of the sulfur from the waste stream previously dumped away (Ludwig Mond). Mond made a small fortune, and soon moved on to another promising technology for soda making.

In 1872 Mond took up a license right of the Solvay process for applications in England from the Belgian chemist Ernest Solvay. The Solvay process is an alternative process to the Leblanc process, originally proposed by the French physicist A. J. Fresnel around 1811 that soda could be manufactured from a brine solution (salt), ammonia, and carbon dioxide. Ever since, it had been tried by many including the James Muspratt who pioneered the Leblanc process in England, but all ended with no vail. While working at the gas works at Charleroi, Belgium in early 1860, Solvay picked up Fresnel's idea and made a few key design changes including the use of a trickling tower to enrich the solution of the ammoniacal salt with carbon dioxide by forcing their contacts, which had been a barrier as the carbon dioxide is hardly soluble. Solvay and his brother, Alfred, built small works to demonstrate the improved process at Couillet, the largest glass making center near Charleroi, Belgium in 1865 (Donnan, 1939). Obviously, there must have been a significant amount of development efforts in order to make it work. Looking at the growing demand for soda and believing that the Solvay process a promising, clean and efficient alternative to the Leblanc process and, however, Mond put the license into work, confident that he would be able to resolve the issues around the technology including ammonia supply, which led to his future development of the Mond Gas process and many more.

Back to England, Mond and his colleague John Brunner formed a partnership, the Brunner Mond & Company, and raised more capital to build a plant at Winnington, Northwich to develop the Solvay process. It was a good start, and the plant started to make soda in 1873. To achieve their expected commercial operation, however, it had taken Mond and his partner additional seven years to implement the system by making the needed changes and modifications due to difficulties with equipment, corrosion and material built-up that plugged the vessels and piping system and so on. Trouble shooting had become a routine daily work. In doing so, Mond had exhibited his chemistry knowledge, engineering skills, and his perseverance.

By 1878, Mond had established the Solvay process into a decent shape, and then moved on to work on a solution for ammonia recovery and additional makeup supply. In Solvay process, ammonia as a catalyst ends up as ammonia chloride, which has to be recovered. By 1880 Mond delivered the milestone by developing a continuous process to recover ammonia from ammonia chloride with milk lime, greatly improving the original Solvay process by reducing the ammonia makeup and putting the whole process on a firm ground (Donnan, 1939). In 1881, Mond and his partner made their soda business public to raise additional capitals in order to expand their operation at Winnington while continuing their efforts to improve the Solvay process. To make up additional ammonia to counter the process loss, Mond in 1886 set up a coke oven at the Winnington site to recover by-product ammonia (Travis, 2018). By this time, the Solvay process had become a formidable competitor to the Leblanc process. The Solvay process manufactured 58,000 tons in 1883, a 480% increase from 1878. While the Leblanc process still dominated the market its annual production increased only 21% during the same period (Lunge, 1884). From early 1890s, the Brunner, Mond and Co. had speeded up its rapid expansion by acquiring other ammonia soda plants that were under stress due to reasons of failing to work technically or working poorly economically. In addition they also developed new soda plants to deploy the Solvay process. By early twentieth century the Solvay process had pretty much replaced the Leblanc process; The Brunner, Mond and Co. also became the largest producer of soda in the world.

It is during this period of time, Mond and his assistants were experimenting methods to artificially make ammonia such as calcining barium carbonate and carbon to form barium cyanide, which releases ammonia when treated with steam. When getting to know a unique way to make ammonia by passing a controlled mixture of air and steam through a hot coal bed where nitrogen in the air ("air-N") would be fixed to form ammonia Mond immediately acted on it to find out more. From coal perspective, in the meantime, Mond also became one of the earliest few, if not the first one, who paid attention to coal-N and its applications in a quantitative manner.

An Unintended Gas Producer

In some way, the creation of the Mond Gas Producer is in fact an unintended consequence of Mond's efforts to search for a practical way to make ammonia, the catalyst to keep his Solvay soda operation afloat. Before building a coke oven for such a purpose, Mond seems to have been purchasing ammonia from coal gas works in Liverpool to make up the balance for the soda operation at Winnington plant (Wisniak, 2006). The ammonia recovered from coal gasworks was not cheap. So was from the coke oven plant. It appears a no brainer that Mond would need a reliable and competitive supply of ammonia in order to secure a sound growth and expansion for his new Solvay process, which had demonstrated promising already by 1878. When Mond began his experiments of fixing Air-N in 1879 the Siemens brothers' Gas Producer had already widely been deployed in the making of steel and glasses. Dowson was in the same year pulling through his producer gas into the 3 BHP Otto Cycle gas engine. With more experiments carried out by passing the mixture of air and steam in a hot coal bed in producer environment, however, Mond and his assistants found out that what was fixed to form ammonia is not Air-N but Coal-N, nitrogen contained in coal. Mond might have thought that he might have hit on something new because of a much higher steam rate and a much lower low temperature adopted in the coal-air–steam system than any of the existing producers at the time. This is somehow reflected in his patent (English Patent 3923) filed in August, 1883.

> The inventor has found that in the combustion of coal in gas producer a low and not a high temperature, as generally supposed, is most conductive to the formation of ammonia from the nitrogen contained in the coal. Nearly the whole of the nitrogen contained in the coal is obtained as ammonia, if the fuel be burned in the presence of steam at a temperature below the point of dissociation. To effect this a limited supply of air, loaded with water spray or steam in large quantity is introduced into the furnace. The resulting gases are richer in hydrogen and have a higher heating power, and the tarry matter is also richer and large in quantity.....
> The point of novelty claimed is the employment of a low temperature of combustion by loading or charging the limited supply of air with so large an excess of steam or water that the temperature of combustion does not exceed a dull red heat.

In subsequent extensive experimentation over the next ten years he and his assistants conducted the coal-air–steam system with different coals optimized operating conditions to maximize the formation of ammonia. In doing so Mond and his assistants must have come to the understanding of the fact that what they needed to extract ammonia from coal would be at least something similar in principle to any of the present producers. What they needed to do additionally, however, is to fine tune the factors such as feedstock, operating conditions, steam condition, capacity, and corresponding parameters to maximize the production and recovery of ammonia from coal gasification. In an ongoing investigation, Mond stated in his President's address to the annual meeting of the Society of Chemical Industry on July 10, 1889, that he and his assistants *"constructed gas producers and absorbing plant of various designs and carried on experiments for a number of years"* (Mond, 1889), indicating the efforts and resources that he had thrown in to recover the small quantity of ammonia, a critical chemical for the Solvay process.

In the meanwhile, Mond knew the value and strong demand for ammonia sulfate as fertilizer through experience. Mond was fully aware that many natural sources such as turfs and others of animal origins (leathers, bones, horns and hoofs etc.) contain a high nitrogen content, for example, turfs have 3+% in mass and animal origins 8–10% or higher. Not like coal, nonetheless, any of these natural materials would be limited in supply. Even though containing nitrogen typically between 0.3 and 2% on a dry weight basis, On the other hand, extracting ammonia from coal still appears the only viable option considering the unlimited supply of it. Mond painted a big picture in his 1889 address that if only 10% of the 150 million tons of coal consumed in England that year were equipped to recover ammonia in the form of ammonia sulfate, it would be possible to deliver a quantity equivalent to all the saltpeter import to the Old World (Europe), which was about 650,000 tons annually. Such a picture looks certainly enticing. The questions remaining are if it makes sense to do it and, if yes, how to make it work, which became the subject that Mond had spent the rest of his life with.

Ammonia sulfate around 1889 commanded a good price of above 12 pounds a metric ton compared to coals only between 6 and 22 shillings a metric ton depending on the quality and location of a coal. It is a significant spread. In order to make sense economically, however, Mond clearly understood that his chance to be successful pretty much hinged on how much more ammonia could be made from coals to produce ammonia sulfate as under such circumstances recovering more or less ammonia downstream with sulfuric acid would impact little the capital expenses of the gas producer system, which would be on a scale that had never seen. During most of the 1880s Mond and his assistants tested many coals mined from Lancashire, Staffordshire and Nottinghamshire areas. In the end, Mond picked a local bituminous slack mined at Nottinghamshire as shown in Fig. 9.2 that contains 1.4% coal-N on a dry basis. It is the same coal that had been used at the Winnington soda works, which was delivered to the works at 6–7 shillings a metric ton. Here is what the number looks like as an estimate about ammonia production.

Mond's early years' experimental investigations achieved only 50% conversion of coal-N to ammonia, and in later years up to 70%, which is significant already under the improved operating conditions of the Mond gas producer (Mond, 1889; Rambush, 1923). With the improved conversion the Mond gas producer would produce up to 50 kg of ammonia sulfate from a ton of the bituminous slack considering a minor loss at the ammonia recovery tower with sulfuric acid. Under such conditions Mond gas producer would need to use an excessive amount of steam 2.5 times of coal rate and a minimal air flow just to maintain a coal bed at the low end of the incandescent temperature range, typically between 900 and 1,200 °C.

To make sense economically, therefore, it appears obvious that the project would have to be on a large scale so to justify the expensive equipment for ammonia production and the subsequent recovery with sulfuric acid. In addition, an unprecedentedly large producer would to be developed understanding that producers in commercial operation at the time, either the Siemens Brothers' Producer, Dowson's or any others, had been designed and made in a small scale, typically processing coal at a rate of several tons daily on a unit basis. Actually, most of the producers were operated on a much smaller scale, consuming less than one ton of coal daily. Considering that one ton of ammonia sulfate would need to gasify 20 tons of the bituminous slack any existing producers would hardly make sense. With the attractive price spread, the cost of 20 tons of coal was only a fraction of the sales of one ton of ammonia sulfate, a promising sign that it might work. Mond moved forward to develop a large scale producer, which eventually became the Mond gas producer.

The early experimental producer that Mond adopted was similar to that of Siemens Brothers', a brick–mortar structure, designed to handle a few hundredweights of coal per day, not more than ten tons of the local bituminous slack daily (Allen, 1908). Again in his 1889 address, Mond provided some details about the producer he adopted in his experimental investigation.

The gas producers which I prefer to use are of rectangular shape, so that a number of them can be put into a row. They are six feet wide and 12 feet long inside. The air is introduced and the ashes removed at the two small sides of the producer which taper toward the middle and are closed at the bottom by a water lute of sufficient depth for the pressure under which the air is forced in, equal to about 4 inches of water. The ashes are taken out from underneath the water, the producers having no grate or fire bars at all. The air enters just above the level of the water through a pipe connected with the blower. These small sides of the producer rest upon cast iron plates lined to a certain height with brickwork, and this brickwork is carried by horizontal cast iron plates above the air entrance. In this way a chamber is formed of triangular shape, one side of which is closed by the ashes, and thus the air is distributed over the whole width of the producer.

Obviously, the grateless design of the Mond gas producer is one of the distinctive features from other producers. It is probably the only way to accommodate such a large producer. Otherwise, a grate would be too big to handle ash discharge, materially and operationally. To carry out the necessary investigations Mond and his assistants set up a large experimental system at the Winnington soda works so that the large amount of fuel gas would be consumed by the soda operation. The system includes equipment of producers, components for ammonia recovery and more accessories etc. Through years' extensive experiments up to 1889 Mond and his assistants had discovered that the operating conditions favoring the production of ammonia would generate a fuel gas rich in hydrogen, a better fuel gas than any other producer gases (Table 8.1, Chap. 8). Encouraged by the strong price of the valuable by-product Mond believed that he could manufacture his fuel gas at a much lower cost than any other producer gases, especially at a large scale. Confident that he was on the right track Mond set out the journey to develop a Mond gas producer system integrated with ammonia recovery.

It seems fair to say that the challenges of both commercial and technical around developing the Mond gas producer system were unprecedented, no less than what Mond had just gone through with the Solvay soda process. This may explain the fact that he changed his mind to move forward, at least prior to his landing on a firm footing with the Mond gas producer, by building a coke oven plant in 1886 to produce the badly needed ammonia for the soda operation at the Winnington site. Among the many technical challenges, the first and foremost comes the operating difficulty associated with the local bituminous slack. Different from the lignite used by Sir Williams and the anthracite by Dowson, the bituminous slack is a caking coal, which means that when subjected to heating to a certain temperature, the coal would go through a mesophase, a transition from solid to liquid, and then become solidified to form a big chunk with the temperature going further higher. In a producer environment, such a transition would create operating difficulties by forming big chunks inside a producer bed. Should the chunk break or crack the subsequently developed crevices or channels would shortcut the passage for air and steam unreacted, not only interrupting the continuous operation but also resulting in a poor quality producer gas because of poor contacts between carbon and air/steam. It could also be hazardous to leave oxygen in the inflammable gases. This is a serious issue that has to be addressed with the producer design. In the meanwhile, the high volatile matter contained with the slack gives off a large amount of tar and oils along with coal gas that needs to be cleaned up before being consumed. This is why Dowson had been avoiding using such coals other than anthracite or coke to avoid such potential operating difficulties, which inevitably attributed to the high cost of Dowson producer gas, almost the same as the coal gas. In addition, the high steam rate used in Mond gas producer causes a significant heat loss to the undecomposed steam, about two third of total steam, leaving the producer along with the resulting producer gas. Steam generation is an energy drain. Clearly, what Mond faced at the time is a whole slew of seemingly unsurmountable issues and challenges. Would Mond be able to overcome them?

Over the next decade, Mond and his assistants got over one issue after another and succeeded in implementing the Mond gas producer system integrated with an ammonia recovery that delivered a quality fuel gas at a very competitive cost. In 1892, Mond upgraded the integrated Mond gas producer system at Winnington site. The new system recovered ammonia to produce ammonia sulfate for sale while sending fuel gas to boilers and gas engines throughout the soda works. The system eventually evolved into a super-sized operation including up to eight Mond gas producers by 1897, gasifying 160 tons of bituminous slack daily and delivering 24 million cubic feet of Mond gas (Humphrey, 1897). Here is a brief schematic description of the integrated Mond gas process.

The integrated Mond gas process at the Winnington soda works was a system including Mond gas producer, superheater, water cleanup, ammonia recovery, gas cooling, and air saturation (Fig. 9.1). During normal operation, the Mond producer receives coal fed through a sealed hopper from the top of the producer. The stream of steam and air superheated at the Superheater, a tube-shell heat exchanger made of wrought iron, passes through the firebars into the producer, which is. The hot Mond gas exiting the producer flows into the Superheater to release much of its heat to superheat the saturate air coming from the Air Saturator at the back end of the process. Such a superheating would help reduce air intake while maintaining a desired high steam rate.

Once further cooled down and scrubbed off carried dust, fines and any other liquid or solids carryovers in the Washer the Mond gas enters into the bottom of the Ammonia Recovery Tower, filled with firebricks to increase contact surface area, where the Mond gas rises up through a cloud of dilute sulfuric acid drops sprayed down from the top of the Tower. Here, ammonia carried in the Mond gas bonds with sulfuric acid to form ammonia sulfate, discharged into a tank. From which the solution is circulated back for additional ammonia recovery while a stream is drawn down to recover ammonia sulfate for sale via additional processes of evaporation, crystallization and separation.

At this point, the Mond gas, almost ammonia-free, enters into the bottom of the Gas Cooling Tower by direct contact through passages provided by woodchips with cold water sprayed down from the top and leaves for storage or consumption by end-users. The heated water discharged at the bottom flows to the top of the Air Saturator and passes its heat and some moisture directly to the rising air before being circulated back for reuse.

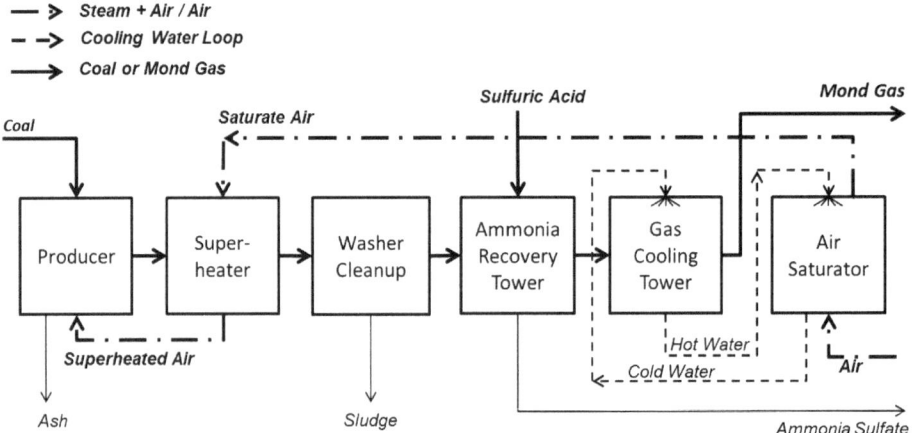

Fig. 9.1 Mond gas producer schematic

The hot saturate air then goes to the Superheater to be superheated before entering into the producer. The Superheater is an added improvement to the upgraded system, which addresses well the heat loss, both latent and sensible, carried away out of producer by the Mond gas and the large amount of unreacted steam to superheat the incoming saturate air. Also, the cooling water loop between the Gas Cooling Tower and the Air Saturator is an effective design to recover additional heat from the Mond gas to warm up and saturate air.

The new Mond gas producer, first commissioned at the Winnington soda works in September 1893, was a cylindrical design with the shell made of wrought iron instead of the previous brick–mortar structure (Fig. 9.2). It was designed to handle 20 tons of the bituminous slack daily, more than double the previous design. With each ton of coal giving off about 150,000 cubic feet of clean Mond gas, one Mond producer would manufacture 3 million cubic feet of Mond gas daily under normal operating conditions. It is a huge leap in a producer design, a scale much larger than any other types of gasifiers including water gas generators and gas producers. By 1895, seven more Mond gas producers of the same size had been added to the site and the total capacity of the Mond gas reached 24 million cubic feet daily for captive use within the soda works (Humphrey, 1897). It was an integrated system with ammonia sulfate as by-product for sale.

The new Mond gas producer bears a few distinctive design features to address some issues arose from the earlier investigations such as handling the caking slack, heat utilization, and others related to scale up of operation etc. It is a double walled cylindrical casing with an inner wall line with firebricks for protection from high temperatures. At about the mid-point, the inner wall starts to taper toward the inside of the producer to form a cone shape to narrow the space as to hold up the shrinking volume of the remaining ash and unreacted carbon. Then, the lower end of the inner wall is connected via a

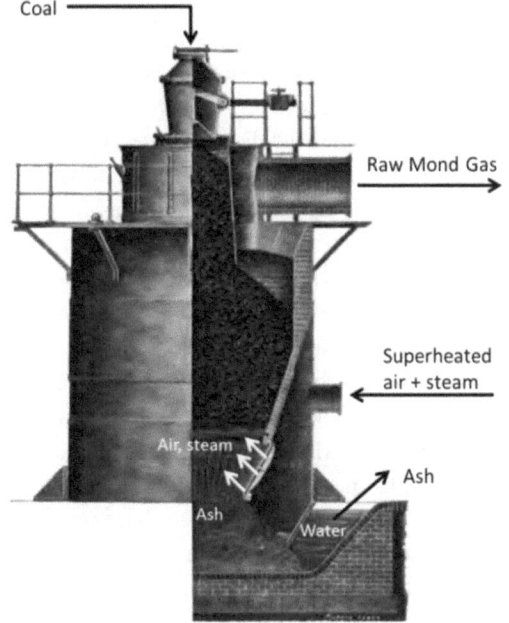

Coal

Raw Mond Gas

Clean Mond Gas Analysis, vol%, Dry

Hydrogen (H)	26.4
Carbonic Oxide (CO)	10.2
Carbonic Acid (CO2)	16.3
Marsh Gas (CH4)	2.5
Nitrogen (N)	44.6
Total	100

Superheated
air + steam

Average Coal Analysis, wt.%, Dry

Nitrogen (N)	1.39
Carbon, total	73.23
Sulfur	1.40
Ash	8.17
VM*	15.81
Total	100

Excluding carbon released above 100 deg C.

Air, steam

Ash

Ash

Water

Fig. 9.2 Mond gas, producer and coal

ring structure to a series of sloped firebars forming a grate, which the other ends of the firebars were hooked up to a smaller ring so to maintain a proper inclination of the grate. The small ring at the bottom of the producer provides an open space that ashes would fall through freely, a feature remained from the earlier design. Through the inclined grate the superheated steam and air, first passing through a jacket or annular space between the double walls, are blown into the producer. Doing so would help cool the inner wall to prolong its life span. Overall, the adopted low reaction temperature and high steam rate would minimize clinkering of coal ash, which is typical to the operations of both producer and water gas generators. During a normal operation, the Mond gas rises up to the space above coal bed around a bell shaped coal hopper, heating up the fresh coal inside the hopper, before existing the producer. Subjected to heating the coal inside the hopper expands and gives off volatile matters as coal gas containing ammonia and some tars. Due to no place to escape the coal gas is forced to flow downward into the hot coalbed where tars in the coal gas get broke down to become fixed in the Mond gas. The bell shaped hopper design addressed the swelling behavior of the caking slack, minimizing its tendency of bridging within the producer bed so that a good contact between coal, coke, air, and steam is maintained so that the Mond gas producer is capable of delivering a large volume of Mond gas of a steady quality.

By 1894, as accounted in the previous section, the Dowson gas fueled gas engines were still hardly competitive over 120 HP against coal fired steam engines. At 120 HP level, the Dowson producer would only be designed to handle about 1.3 tons of anthracite daily, indeed a small unit compared with the Mond gas producer. The Dowson producer, nonetheless, did well at a small scale because it occupies much less space, requires less labor, and is easy to maintain. The Dowson producer also uses much less coal during a hot stand-by, and quick to return to normal operation. The Dowson producer was in general favored by small shops and owners who weigh more on the convenience than gas price. At large scale, however, these benefits tend to diminish and the cost of fuel gas becomes important in order to compete with steam engines. Notwithstanding the progress that Mond had made with his integrated Mond gas process so far it is a complicated system including those necessary components such as for ammonia recovery and heat recovery etc. It would cost a lot more to develop, build, operate and maintain. In addition, another potential risk could well be that the large amount of the Mond producer gas could become a liability once the plant would have been built. The question then becomes obvious, that is, would Mond be able to overcome the price hurdle to compete with steam engines. Such a challenge would certainly appear not small at all in a sense that he would have to deliver the large amount of fuel gas at a price low enough, on the one hand, and to secure enough off taking customers, on the other. Considering the scattering market conditions back then it certainly seems a daunting task at least. Concerning gas engines, makers were reluctant to make gas engine larger than 120 HP because of the perception that the expensive producer gas, though cheaper than coal gas, would not be competitive at all against steam engine. Although the scale economy of a larger plant would help reduce the unit cost of the Mond gas securing enough off-taking clients to absorb the large quantity of Mond gas would, however, be another story. According to Humphrey, however, the integrated Mond gas process at the Winnington soda works proved promising and would be able to deliver a positive result.

Humphrey? Yes, it is Herbert Alfred Humphrey who invented the Humphrey pump, but that is about a decade later. Humphrey joined the Brunner, Mond and Co. around 1890 to assist Mond to assess the performance of the integrated Mond gas process, and the applications of the Mond gas with boilers, gas engines, and their economics at the Winnington site. He made a significant contribution to the commercial deployment of the Mond gas process. In his presentation to the annual meeting of the Institution of Civil Engineers on March 16, 1897, Humphrey provided some detail accounts of his findings from the investigations (Humphrey, 1897).

By design, the new integrated Mond gas process plant had eight Mond producers, each rated at 20 tons per day. At a full capacity, the plant would gasify 160 tons of bituminous slack daily and produce 24 million cubic feet of Mond gas. Equipment is arranged in two groups with each group consisting four bundles of Mond gas producers integrated with an Ammonia Recovery Tower, a Gas Cooler, and an Air Saturator. In each of the bundles, each Mond gas producer has a dedicated Superheater and a Washer. During a normal operation, producer gas from four Mond producers would be routed into one set of the downstream unit for ammonia recovery. According to Humphrey, some of the old brick–mortar producers were still in working condition and were put into operation when necessary. The investigative runs took place on one group of equipment in the summer of 1895, Humphrey carried out three cases with different coal feed rates at 62, 84 and 98 ton daily, respectively corresponding to below, on and above capacity with four producers in operation, which is a typical methodology to test out the resilience of a process system. Conditions of steam and air were optimized to favor the formation of ammonia. During the three day runs for each case the Mond Producer handled well the bituminous slack and produced the Mond gas of a consistent quality. As would be expected, the below capacity case produced the best performance because gasification reactions have longer time to proceed. Humphrey also found out that absorption of ammonia by sulfuric acid at the Ammonia Recovery Tower was a restricting factor. On average each ton of ammonia sulfate requires 23 tons of slack as feedstock to producers and additional 5.5 tons of the slack for generating the needed steam, totaling 28.5 tons. Humphrey estimated the cost of the Mond gas based on prices in 1896, bituminous slack purchased at 6s 2d a ton and ammonia sulfate sold at £7 4s 6d a ton, which the latter dropped significantly compared to 1889. The sale of the ammonia sulfate, however, still covers 57% of the total operating cost of the integrated plant including materials (coal, sulfuric acid and lubricant etc.), wages, operating and repairs and utilities. The cost of 1,000 cubic feet of Mond gas was 0.35d, a small fraction of Dowson gas or coal gas at the time. Humphrey further made a comparison without ammonia recovery, and the cost went up by 44% but stood at only 20% of the Dowson gas on the same basis noticing that Dowson gas paid more than 21s for a ton of anthracite.

Humphrey's findings certainly confirmed Mond's belief that his new design would be able to overcome the hurdles presented by the coals of low quality, the large scale operation, and the fuel gas of a good quality while maximizing the formation of ammonia to make ammonia sulfate, a valuable by-product that helps offset the cost of the Mond gas. Encouraged by the results Humphrey moved on to test the Mond gas for different applications with such as coal fired boilers, iron and steel furnaces, gas engines and heating for evaporation, and so on around the soda works. The Winnington soda works turns out to be a very valuable platform for the development of the Mond gas producer because the works burns a large amount of coal in boilers to generate steam to run engines for driving power and to feed heating processes for evaporation needs, which could be easily switched to gas firing. It is noteworthy that changing coal fired furnaces to gas firing had contributed to significant savings with labor and repairs around the soda works operation, but the fuel saving was insignificant because the soda works was running around the clock so a standby waste of coal firing was of less a concern. The Mond gas also found itself a suitable fuel for steel and iron making as well. Then Humphrey turned his focus on gas engines which were used to provide lots of mechanical power throughout the soda works.

As early as in July 1894, Humphrey ran a 25 HP Otto gas engine for 2 h with the Mond gas. The gas engine, also made by Crossley Brothers, achieved a thermal efficiency of 23.8% with the Mond gas. What is interesting is that the gas engine drove a 20 kW dynamo generating a direct current (DC) at 100 V, which was consumed by a resistance wire. Then in April 1896, Humphrey ran the same gas engine to drive a 30 kW dynamo, and used DC electricity to power both incandescent lamps and arc lamps for lighting up the soda works. The saving compared to using coal gas illumination is significant. The soda works, therefore, installed a new Otto gas engine of 150 HP coupled with a 75 kW dynamo made by Siemens to generate more DC for lighting around the works. The gas engine is of a two cylinder, end to end type running at 160 RPM. This became the first use of a large gas engine ever integrated with a dynamo for electricity generation in England. In essence, this demonstration can be deemed as the earliest integrated gasification with a gas engine and dynamo for electricity generation, the early version of the integrated gasification combined cycle (IGCC), which is still under development today. Encouraged by such a success and the viable low cost Mond gas, makers of gas engines began to make large engines of 500 HP, 1,000 HP and even larger. The Mond gas had subsequently made its way to break the bottleneck that Dowson gas had faced for quite a while to compete into the large steam engine market, which seems limitless from the perspective of generating electricity back then. The excitement and expectation at the time can be sensed from the following statement made by Humphrey (1897).

The atmosphere of London would be relieved of the smoke which makes a London fog so objectionable, for factory owners would be supplied with power at a cost which even the Niagara Falls Power Company of America cannot reach. Also the expenditure of England in nitrogenous compounds or fertilizing agents, amounting to about £2,000,000 per annum, would, as the system of gas-producers became general, be changed to an annual income arising from the sale of the surplus sulphate of ammonia in foreign markets.

Unfortunately, it appears that further deployments of the integrated Mond gas process and gas engine coupled with dynamo for electricity generation on an industrial scale did not take place until 1901 when Mond set up the Mond Gas Company in Staffordshire, England.

Mond had shifted his priority to other areas that he believed much more important for deploying the Mond gas.

Mond Gas, Fuelcell and Nickel

The incandescent lamp that Edison invented in 1879 essentially began the ending of the coal gas age for illumination. Like what happened sixty years ago when coal gas replaced candles and oil lamps, the convenience and cleanness of the incandescent lamps attracted publics early on. In a panic, coal gas companies were forced to slash their gas prices to retain their customers in order to stay in the game. What the Shanghai Gas Co. did to reduce its gas prices back in 1881 is one of the many examples. By 1880, technologies for generating electricity had existed for decades: gas engines running satisfactorily with coal gas or producer gas, gas engines driving dynamos successfully by direct coupling or via a belt to force electrons to flow in one direction that could be used to light up arc lamps. In addition, there was a variety of them to choose from. From gasification perspective, one of the major obstacles is the high cost of coal gas or producer gas, which attributed to the high cost of electricity. The reliability issue of the incandescent bulbs itself is another to worry about. The trend, however, had set in that electric lighting would replace coal gas illumination. It is only a matter of time. Mond saw this trend as an opportunity without a ceiling. He believed that an idea that he was about to place his hands on would make a wonderful technology to generate unlimited amount of cheap electricity, which would help him open up a huge market for the large amount of the Mond gas that Mond would most likely have trouble to dispose of.

Fuel Cell for Electrons

Back to early 1880s when Mond and his assistants found out that their producer, when maximized for ammonia formation, would result in a producer gas rich in hydrogen, about 40% higher than the Dowson gas. This must have caught Mond's attention as he pulled off an invention from the shelf that William R. Grove discovered in 1839. The invention, called the Grove cell or gas voltaic battery, is to turn hydrogen directly into electricity. Although Grove and many others had tried the idea it did not go anywhere. Viewing it from the perspective of Mond gas, Mond believed that this idea, if developed, could open up a huge and insatiable demand for the Mond gas. In 1884 Mond hired Carl Langer, a German chemist, into his newly opened laboratory in London to tackle the subject.

In concluding his president address about the formation of ammonia from coal in 1889 Mond provided a brief account of two more inventions as a result of his efforts in developing the Mond gas producer.

> Before leaving my subject, I will, if you will allow me, give you in a few words a description of two other inventions which have been the outcome of this research. While looking one day at the beautiful, almost colorless, flame of the producer gas burning under one of our boilers, it occurred to me that a gas so rich in hydrogen might be turned to better use, and that it might be possible to convert it direct into electricity by means of a gas battery.

What Mond and Langer did first is to reproduce what Grove did to get familiar with the Grove gas battery. Once getting over the learning curve they started to tackle the flooding issue by trying different materials, electrolytes, and designs during the time between 1885 and 1889, which led them to build the first dry gas battery or fuelcell in today's term

(Langer, 1888; Wisniak, 2006). Such a fuelcell has remained by large the prototype of fuelcell that is still under development to this day.

Their first gas battery was a seven cell one. Mond and Langer first tested the battery with hydrogen supplied into one chamber and air to the other the battery generated a current of 2 A of 5 V. What is interesting is next when Mond and Carl introduced coal gas, the Mond gas, to replace hydrogen in the fuelcell. The Mond gas as it was worked the same as hydrogen.

Mond and Langer seem not surprised by the fact that the Mond gas produced practically the same result as hydrogen. They probably hoped that carbonic oxide in the Mond gas would act similarly to hydrogen by being oxidized to carbonic acid, which was actually proposed a few decades later as a fuel gas because carbonic oxide and oxygen is just another pair of redox reaction. It would be great if carbonic oxide works the same as hydrogen remembering that the Mond gas contains more than 10% carbonic oxide in addition to 26% hydrogen. Unexpectedly, however, the current generated from the fuel cell soon dwindled away. After a while, Mond and Langer found that the platinum black on the fuel side was covered with carbon particles and lost adsorption power for gases. Carbonic oxide and hydrocarbons such as marsh gas and ethane etc. become cracked over the platinum black, which poisoned the fuelcell. There rises the need to remove carbonic oxide and hydrocarbons from the Mond gas, which is the second invention that Mond mentioned in his 1889 address.

Gas separation technologies that are widely practiced in modern industries are not existent back in the 1880s. If there was, it would still be in laboratories working on a small scale. Mond and Langer found out that passing the Mond gas along with steam through some metals or their oxides was an effective way to remove carbonic oxide and hydrocarbons. This finding might explain the encounter that carbonic oxide was cracked into carbonic acid and carbon and other hydrocarbons into hydrogen and carbon over the platinum black. Upon screening, they landed on nickel and cobalt which give the best result by destroying those components almost completely by steam at a temperature between 350 and 450 °C. The Mond gas exiting the producer seems to suit such conditions well. To execute it Mond and Langer placed a porous materials such as pumice stone impregnated with nickel or cobalt in a retort, which is heated externally only at the start. Once the hot Mond gas flows through the retort the reactions would produce enough heat to self-sustain. The treated Mond gas, now free of carbonic oxide and hydrocarbons, has a hydrogen content increased to 36–40% from 25% and become suitable for the fuel cell. The success at the laboratory scale made Mond confident that "*we have no doubt as to its complete success…at a very small cost*" at a future large scale. Obviously, the expectation appeared high at the time that coal gas fed fuel cell would generate electricity at a much cheaper cost than any dynamo driven by either a steam engine or a gas engine. The fuel cell technology would soon open up a huge demand for the Mond gas process.

Unfortunately, the ongoing development work of fuel cell was interrupted due to other more pressing issues around his soda business.

The Magic Carbonic Oxide

By mid 1880s, the Leblanc process still had a monopoly of calcium hypochlorite, a bleaching powder with a strong demand by the industries of textile and paper makings. Mond decided to wait no more and gave it priority to upgrade his ammonia recovery process that had been in use but left the chloride as a waste in the form of calcium chloride. After surveying potential ideas for a better outcome Mond this time decided to develop a one-step process to recover ammonia and chlorine from ammonia chloride out of the Solvay tower. In the upgraded process Mond and Langer first crystallize ammonia chloride by cooling the residual salts solution, and then heat up the solid ammonia chloride in a separate vessel to about 350 °C so that ammonia chloride would decompose into vapors of ammonia and hydrogen chloride, which enter into another vessel packed with pellets of nickel oxide maintained at about 400 °C where hydrogen chloride becomes selectively absorbed by metal oxide to form nickel chloride, and ammonia passes through to be recovered for recycle. In next operation, the spent nickel oxide would be subject to regeneration. Depending on the use of hot air or steam chloride would be released as chlorine or hydrogen chloride as by-product. It is a flexible process that had proved to work well in the laboratory, and based on which a patent was filed in 1886 (Mond, 1886). When moving to Winnington site to carry out a large scale testing, however, an unexpected took place, which led to another significant discovery.

Being well aware of the highly corrosive nature of hydrogen chloride vapor Mond and Langer had to line up the inside of the testing vessels with corrosion resistant materials such as ceramic material, and used nickel valves in the line where the vapor passes through. In an actual testing, a purging between the absorption and the regeneration would be necessary to clear up the line and absorption vessel. After some time of the testing, however, Mond and Langer found that the valves became leaking. Upon examinations, they found that the bed of nickel pellets became disintegrated, and both inside of the valves and the nickel pellets were covered with a carbon black. Confident that there had been no change in how they operated the system except that the purging gas used in the laboratory back in London was bottled nitrogen, but an off gas from the limestone operation at the field testing, which was believed to be carbonic acid and should act the same as nitrogen. Upon a closer analysis, however, a small amount of carbonic oxide was found in the off gas. Reminiscent of their experience with platinum black and platinum in their previous investigations of the fuel cell and hydrogen recovery by using nickel and cobalt, Mond found that this carbonic oxide is a very interesting gas that has a strong affinity with metals and their oxides. Mond directed Langer to find out more about this carbonic oxide and its behavior over metals remembering that carbonic oxide is a major component of the Mond gas.

Back to the laboratory in London, Langer lined up the instrument by placing a fine nickel powder in a glass tube, passing carbonic oxide over the nickel powder heated to different temperatures each time. To avoid the emission of carbonic oxide a Bunsen lamp was positioned at the end of the line so that any remaining carbonic oxide would be destroyed. In each run, the nickel powder was treated at a temperature in a carbonic oxide atmosphere for some time and then cooled down to room temperature. Each time, the resulting nickel powder was analyzed. On an occasion, Langer accidentally observed some peculiar phenomena at the Bunsen lamp during the cooling process. When the glass tube cooled to about 150 °C the flame of the Bunsen lamp became luminous with increased brightness and then changed to greenish yellow once temperature dropped further to below 100 °C. After some wild thoughts, Langer heated up the glass tube where gas flow continued and suddenly the glass at the Bunsen lamp became coated with a bright metallic mirror. In the meantime, the lamp lost its luminosity. The coating was found to be nickel! The nickel must have formed a kind of compound with carbonic oxide and then decompose to release nickel upon heating at the flame. Further isolation and examination of the compound found that it is $Ni(CO)_4$, a liquid compound that boils at 43 °C and freezes at 23 °C, a completely new compound named nickel carbon oxide or nickel tetra carbonyl. Mond and Langer published their finding in 1890 with the Journal of the Chemical Society. While continuing the investigation, Mond was contemplating how to put this finding into use. If carbonic oxide could be used to separate nickel from other metals such as cobalt, iron and copper etc. it might be well possible to put the Mond gas into use to develop a completely new process to make nickel of high quality and the cost to do it would be low because carbonic oxide possesses such a strong affinity with nickel at such a low temperature. It would be a huge win. Nickel, the fifth most common element on earth, is a lustrous, silver-white metal; it is hard, ductile, malleable, and can take a high polish that is a highly valuable commodity to make steel special.

Mond jumped right into the action.

Mond Gas and Nickel Refining

Although the metallurgical procedures to process nickel ores had been pretty much established in the 1880s the nickel of a high purity, however, was not easy to make, which had been a subject of researches and inventions around mid-1880s (Abel, 1884; Readman, 1883). With more resources poured into the laboratory in London, Mond put Langer in charge in 1892 to construct and manage a pilot plant at the Wiggin's Nickel Works at Smethwick, western Birmingham, to demonstrate the process developed thus far at a laboratory scale. Then with several more years of patience, troubleshooting and systematic work, the famous Mond Nickel Process, capable of manufacturing nickel of as high as 99.9% purity, was born. Here is how it works and how the Lowe's water gas, plays magically to refine the nickel from other metals.

The Mond nickel refining process takes place in three steps, reduction, vaporization, and decomposition (Fig. 9.3). Nickel matte, a concentrated nickel oxide containing some oxides of cobalt, iron and copper etc., is introduced into the reduction chamber heated to

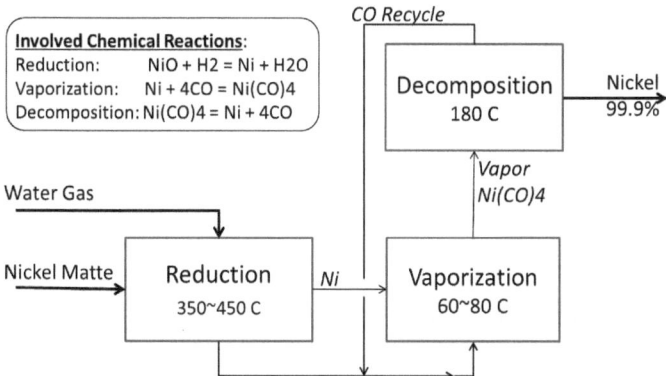

Fig. 9.3 Schematic Mond nickel refining process

a temperature between 350 and 450 °C where the nickel matte meets with a water gas and is primarily reduced by hydrogen therein to nickel; Obviously, Mond used Lowe's water gas instead of the Mond gas because of its much higher concentration of hydrogen and carbonic oxide. Then the reduced nickel cools down to a temperature not exceeding 80 °C when being transferred airtight to the next chamber where carbonic oxide in the remaining water gas (now rich in carbonic oxide) selectively reacts with nickel to form nickel tetra carbonyl as vapor, which is carried off in gas phase to the final chamber heated to 180 °C for decomposition. The decomposed nickel grows on nickel pellets giving high purity product of as pure as 99.9% while the released carbonic oxide is sent back to the vaporization chamber for reuse. The unique and strong chemical affinity and high selectivity possessed by carbonic oxide over nickel make this process a charm, so simple while consuming little energy.

Encouraged by the outcome Mond stepped into the mining business by purchasing the nickel-ore properties near Sudbury in Ontario, Canada, and developed related infrastructures to pretreat the nickel ore into nickel matte, which was shipped back to England for refining. In 1900 Mond established the Mond Nickel Company to manage the nickel business under which the Mond Nickel Works was set up in Clydach near Swansea, the metallurgy center, to refine the nickel matte. The Works was commissioned in 1902, and produced 3,000 tons of nickel of 99.9% purity by 1910. By 1939 the Mond process produced a third of the nickel consumed worldwide. After the death of Mond the Nickel Company merged in 1928 with the International Nickel Company ("INCO"), an American company. In 1932 INCO registered the trademark of "Inconel" in the US, a family of special alloys made of nickel and chromium previously invented by the Mond Nickel Company's subsidiary, Henry Wiggin & Co. in Hereford, England. The Inconel, a family of super alloys, extends the function of steel because of its superior performance by being corrosion and oxidation resistant under high temperatures and high pressures. Early Inconel demand coincided with the development of jet engines. After the 1950s, the industrial expansions in space and aviation, automotive, power generation, chemical, petrochemical and refining etc. prompted an urgent need for better materials, which created a significant development to expand the Inconel super alloys. Some of the super alloy materials have been widely deployed in modern gasification facilities such as servicing the high purity oxygen lines and other highly corrosive and erosive environment including radiant coolers. Such advancement in materials in turn helps greatly the modern gasification technology move to another level.

Mond Gas, Industrial Gas and Electricity

Having upgraded the old ammonia recovery process with the new one to recycle both ammonia and chlorine around 1887, though not perfect at the time, Mond had essentially ended the monopoly of the bleaching powder, the last straw for the Leblanc process. The simple and environmentally friendly Solvay soda process had, therefore, forced the Leblanc associated businesses to consolidate in 1890 to form the United Alkali Co. in order to stay afloat. By 1900, the Brunner, Mond, and Co. manufactured 90% of soda products worldwide with its business present in many countries and regions.

Although the development of the fuel cell appears to fall short of Mond's expectation, possibly prevented by other priorities, Mond by the early 1890s must have realized that he would face mounting challenges that had to be resolved. The technical challenges such as the Mond gas purification, electrode materials, and difficulties related to the scale up etc. would at least take additional time and resources to tackle. In retrospect the framework and the prototype of the fuel cell that Mond and Langer developed had become the principle or guideline for many in the following decades in their efforts to advance the fuel cell technology that would work and make economic sense. Such challenges have yet remained so as of today except for a few niche cases where economics might be less a priority such as the onboard hydrogen fuel cell for the Apollo Space missions that NASA launched in the 1960s. In the deployments, the water discharged from the fuel cell operation was used as drinking water on-board, which would otherwise go wasted for other projects on the earth. Another example is the recent efforts to deliver hydrogen fuel cell cars, trucks and buses to fight the climate change as part of the energy transition. The Toyota Milai and the Hyundai NEXO etc. are a manifestation of the on-going progress about the fuel cell technology.

To be more effective, Mond appears to shift his efforts back to conventional applications of the Mond gas with gas engines, furnace, boiler heating, and other metallurgical uses etc.

Early Industrial Gas Network

In 1901, Mond set up the Mond Power Gas Corporation at Stockton-On-Tees, to own and manage all Mond gas related intellectual properties, engineering services and equipment fabrication in order to grow his Mond gas business. In the same year, Mond also incorporated *the South Staffordshire Gas Company* under a license granted by the Parliament Act in 1901, 91 years after Winsor did. This time, however, the objective of the company was to manufacture and distribute the Mond gas to a neighboring industrial network for uses other than illumination, which marks the first example of industrial gas utility. As stipulated under the Act, the purpose of the company is "*to manufacture, supply, sell, and distribute within the South Staffordshire and East Worcestershire district the producer gas commonly known as Mond gas and any development of Mond gas approved by the Board of Trade and any other producer gas so approved for the purposes of motive or other power, heating, or any other purpose to which such gases can be supplied (except that such gases shall not be supplied or used for the purposes of illumination), and to manufacture, sell and deal in sulfate of ammonia and any other bye products or residual products of the said gases, and generally to execute the powers and purposes of this Act*" (Keen, 1901). In particular, the Act spelled out that the company was prohibited to supply its Mond gas or any other gases, directly or indirectly, for the purpose of illumination and the subsequent liabilities should any such a violation take place. Nevertheless, it would be the first ever and the largest undertaking to distribute and sell producer gas to an industrial network.

Under the Act, the industrial network covers a jurisdiction of about 123 miles^2 of area, from the north boundary of Birmingham to Wolverhampton and from Pelsall to Stourbridge, where about 640,000 inhabitants and more than 2,200 shops, large or small, resided. There were also big operations from coal mining, porcelain making, and limestone production that were routinely in need of power for both driving and heating, estimated at 300,000 HP equivalent of the Mond gas, which, if realized, would aggregate to about 3,000 tons of coal daily to be gasified. The prices of the Mond gas that the company could sell were, however, dictated by the Act. The charge to its clients or customers shall not exceed 3d per 1,000 cubic feet of the Mond gas if the quantity sold shall be no less than 4 mm cubic feet during a period of 13 weeks. Otherwise, the charge shall not exceed 4d per 1,000 cubic feet. Such a huge potential demand would present Mond a significant opportunity to grow the Mond gas business.

Similar to what GLCC did back then, Mond sited his Mond gas plant at the center of the territory, Tipton, which is conveniently situated next to the Stour Valley Railway Line and the Dixon's Canal Branch, and just to the east of the Tipton Gas Works commissioned in 1882 providing coal gas to the surroundings. In its first phase, Mond deployed eight producers, each gasifying 20 tons of local coal daily delivered onsite by barge through the canal. The construction opened ground in 1902 and was completed three years later. To minimize the overall risks of the project, the design and layout of the first phase of the Mond gas plant appear to just follow what was built into the old Winnington plant, gasifying 160 tons of coal daily integrated with co-producing ammonia sulfate for sale. The project also used the same local bituminous slack as feedstock. What's worth noting though is that the Mond gas plant at the Winnington consumes most of the generated Mond gas onsite by the Solvay soda operation and with only a small portion being sold over the fence to an electric company nearby. The Mond gas plant at Tipton would act essentially as a central station to produce the Mond gas. Except for a small quantity for onsite auxiliary use, most of the Mond gas would be transmitted through a network of pipelines that would eventually cover the territory of 123 miles^2 area to industrial users. There arise quite a few challenges that Mond had to face. One example is that the Mond gas or producer gas in general is a "heavier" gas having a specific density of 0.86 lb./cf, due to its high nitrogen content than the traditional coal gas of 0.42 lb./cf, and pushes it through the distribution network requires a much higher head pressure. Although coal gas had been widely distributed via a well-established practice, if not standardized one, reengineering the pipeline network for a producer gas would be necessary to address the new challenges. Here is a brief description of the early Industrial Mond gas network.

The Tipton plant, sitting on a 40 acre lot, had a few sections as shown in Fig. 9.4. Once delivered onsite coal is offloaded directly to coal bunkers, and then conveyed automatically and distributed to storage bins situated over each of the two groups of producers. Each bin holds 40 tons of bituminous slack. An electric motor of 5HP drives the conveyers.

In short, the Mond gas making section integrated with ammonia recovery and its operation is similar to the layout developed at the Winnington site. Processes and equipment to prepare and compress the Mond gas before entering into gas main are the added features to the Tipton plant. The Mond gas, after leaving the Integrated Mond Gas Plant, would be subject to additional clean-ups through two large centrifugal fans, each driven by a 45 HP motor, and two scrubbers to remove remaining tars before going to gas meters of a rotary type. Last, the metered clean Mond gas would be compressed by three 450

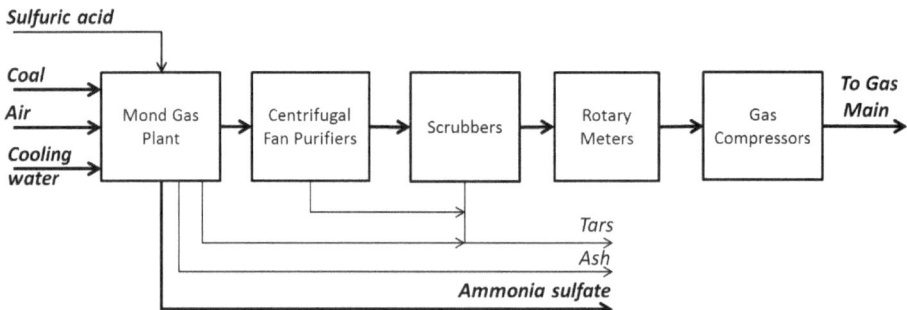

Fig. 9.4 Schematic of the centralized industrial Mond gas network

HP compressors to 10 psi to feed into a gas main. Each of the compressors is capable of processing 24 mm cubic feet of the Mond gas daily. Under a normal operation scenario, two of the compressors are in operation and one spare; so are the gas meters.

In modern day practice, a natural gas pipeline operated at a pressure of up to 2,000 psi has been well established. This is not the case back in 1900. The Mond gas pipeline designed at 10 psi was the first of its kind and had never been experienced. The traditional coal gas infrastructure had been designed typically at a slightly positive pressure, a few inches of water pressure. To design a transmission system of 10 psi for the "heavy" producer gas, gas mains would have to be constructed with steel material, instead of cast iron used for the traditional coal gas system, and coated internally with a layer of asphalt, which was further enhanced with canvas. The asphalt could be manufactured from tars. Gas mains of 36″ in diameter was used at the central station and reduced to 21″ at the end of the distribution. The delivered gas pressure at the end of the distribution shall be maintained at 5 psi, high enough, and then let down with a valve to different pressures that would suit each individual end-user.

The initial plan for the gas main was to run north from the gate of the central station to Toll End and then Ocker Hill where it splits into two branches, one branch heading northwest to Wolverhampton via Bilston and the other toward the northeast to Walsall by passing Wednesbury. By about 1907–1908, there were 13 miles of gas mains and branches being laid. The number had stretched to 37.5 miles by the early 1920, connecting 161+ works that used the Mond gas for a variety of purposes including metallurgy and heating operations, and driving gas engines etc.

From the perspective of performance, the central station of the industrial gas network during the year of 1919 produced 4.4 billion cubic feet of Mond gas out of about 29,500 tons of coal consumed, and recovered 1,280 tons of ammonia sulfate by-product representing a yield of 4.3% by weight of coal consumed by gasification. About 89% of the Mond gas was sold to different industrial end-users. Prior to WWI Mond sold his gas at prices much lower than what was stipulated by the Parliament Act 1901; each 1,000 cubic feet of the Mond gas charged 1.5–2.75d for gas engine use and 1.5–2d for heating purpose. During the war, however, prices of coals, materials, and labor and so on had gone up significantly. Mond was able to increase the prices as well by charging 8.8–10d for engine use and 8.8–9.3d for heating for each 1,000 cubic feet of the Mond gas in 1919. Assisted by ammonia sulfate that was still selling strong, the 1919 was a still profitable year at *the South Staffordshire Gas Company*. The sale of the by-product accounted for about 12% of total sales. Here is a peek into the sales versus expenses for the operation of year 1919 (Table 9.1).

It seems too sketchy to talk about the economics about this operation. Considering the total capital expenses of 548,030 pounds that had incurred by the year end of 1919, nevertheless, the net number may well tell a decent performance already considering the timing. The significance of the Mond gas as an industrial gas distributed radiantly through a pipeline network, however, may hardly be measured only by its financial success; more

Table 9.1 Operating margin of the Mond gas plant with ammonia sulfate recovery

Sales	Pound	Expenses	Pound
Mond gas	130,609	Manufacturing costs (coal, payroll, O&M and materials)	119,370
Meter rentals	1,020	Distribution cost (payroll, O&M, materials and lab work)	9,026
By-product	19,115	Rentals, rates, taxes and insurance	3,122
Others	176	Management costs (payroll, and general expenses)	3,545
Total	*150,920*		*135,063*
Net	15,857		

Source Producer Gas, E. Dowson and A. T. Larter, 1920

appropriately from a technology perspective, it ought to be measured by the breakthroughs that Mond had made with his decades of unwavering commitments to advance gas producer to such a large scale, a scale that was far beyond what was achieved by Siemens brothers and Dowson, and to explore the cheap coals of bituminous slack to deliver a quality gas fuel in a large quantity and at a low price, which in return attracted more industrial users. Of course, the unique position and value of ammonia sulfate at times played into this undertaking in many ways, magnetically if not magically. As Dowson put it plainly that *"The late Dr. Mond and his associates have done valuable pioneer work in proving what can be accomplished, and others will doubtless profit by the experience gained* (Larter, 1920)." has leapfrogged the gasification technology into the twentieth century to play more.

Mond Gas—The Clean Fuel

Before WWI, traditional coal gas was sold above 2s for every 1,000 cubic feet, but Mond gas at as low as 1.5d or 6–7.5d equivalent to coal gas, which is typically 4–5 times the heating value of the Mond gas. The low cost Mond gas had created significant demand in the industrial sectors for a variety of purposes, driving gas engines for mechanical power and electricity generation, firing up boilers, heating furnaces, cooking, and many more. Undoubtedly, this had pushed the traditional coal gas business to the corner.

Firing Power of the Clean Fuel Gas

Around the time 1870s, England society as a whole were facing another dilemma, that is the poor air quality resulted from coal burning especially in cities and towns where extensive industrial activities had existed for a long had become a public nuisance as hundreds more chimneys stood up high and constantly spewed out black smoke, a choking, dusty, and greasy smoke. Compared to what happened during the reign of King Edward I in the early fourteenth century the public nuisance in the late nineteenth century had taken a whole new dimension that had caused increasing public outcries and complaints, especially in major industrial cities such as Birmingham, Manchester, Glasgow, Leeds etc. It was impossible to stop, ban, or even reduce the use of coals; coals had become part of the public lives. Under such an environment, there seems an obvious opportunity for the well-established coal gas business to expand into the industrial sector to replace the direct coal firing as a quick solution to remove the black smoke. Unfortunately, the problem is the high cost of coal gas that industry owners had no margin to absorb. The coal gas industry started to look aggressively into technologies and commercial approaches as well to improve their operation. In Birmingham, for example, the city consolidated five gasworks including the Winsor Street Gas Works, the Birmingham Gas Light Co., and the South Staffordshire Gas Light Co. into the Birmingham Corporation in 1875, and launched a plan to revamp its operation by building new coal gas storage tanks to reduce gas leaks and upgrading its coal gas making process with new technologies such as the regenerative retort, the water gas process, and the Woodall and Duckham vertical retort. All these helped improve the coal gas making operation while reducing the cost of the coal gas. To facilitate the shift from coal to coal gas firing, the gas department of Birmingham also set up incentives to encourage public and industrial owners to use more coal gas or to switch to coal gas, especially to upgrade steam engines to gas engines or to modify coal fired boilers with available coal gas. The gas department also offered expert advice, free of charge, to whoever would consider such a switchover. The advantages of gaseous fuel as a clean fuel over coal had been well recognized. Industrial users would welcome such a change to replace their dirty, laborious, and maintenance extensive coal fired boiler with a clean, affordable, and low wear and tear gaseous fuel. By 1897, the Birmingham Corp. supplied its coal gas to 1,600 gas engines via its 618 miles long network, and four years later, the number increased to 2,408. Of the 2,408 engines, 1,300 engines were under 6 HP, 1,000 between 8 and 14 HP, and only 100 of 30 HP or larger. Similar happened to towns such as West Bromwich, Walsall, Smethwick, and Wolverhapmton etc., which is what the South Staffordshire Mond Gas Co. would cover (Keen, 1901). Some of these small gas engines were adopted to provide mechanical power and some were coupled with dynamos to generate electricity for lighting.

With the coal gas market moving toward gas engines, industrial heating, residential heating and cooking, away from the traditional illumination use, the daily pattern of coal gas use had changed as well in all cities and towns in Britain. Previously in Birmingham, coal gas demand peaked between sunset and sunrise. Now, coal gas demand during day time from 6 am to 6 pm had become the same as that between 6 pm and 6 am, basically around the clock (Londoner, 1912).

Encouraged by the demonstrations of the integrated Mond gas process, gas engine makers like Crossley & Bro., Cockerill, Westinghouse, Nurnberg, Koerting, and Premier started to develop and offer large engines designed for gaseous fuel applications. Before 1908 there were about 39 large gas engines worldwide running or under construction with Mond gas among which nine were 500 HP, ten 650 HP, four 750 HP, and six 2,500 HP, which is a significant advancement from what Dowson had achieved. As a result, gas engines had penetrated further into the market so far dominated by steam engines for more than a century (Fig. 9.5).

In addition, the Mond gas producers had been used to fire up boilers for steam generation. In Manchester, for example, a local court stepped in to settle the public disputes over the black smoke emitted from Trafford Power and Light Supply Co. and W. T. Glover and Co near Trafford Park. Both the companies finally installed the Mond gas process to fire up their boilers. The representative of the former company, who ran two 1,000 HP boilers to generate steam for generating electricity, promised to court in January 1902 that the Mond producers under installation would soon replace the coal firing in the two steam boilers, and "*in a fortnight's time the Mond gas would be in use for the purpose of firing up, and there would be no black smoke*" (Co, 1903).

Mond Power and Gas Corporation also sold its gas producers overseas such as to the Taikoo Dockyard in Hong Kong before 1908, one of the major dockyards in Hong Kong and capable of handling the largest ship repairs in the world at the time. The Mond gas was used to drive gas engines and other miscellaneous uses for heating. The Dockyard was developed by John Swire and Sons, Hong Kong, Ltd. between 1902 and 1907 where there installed four gas engines of Cockerill model, two 1,100 HP and two 500 HP coupled with dynamos with a capacity of generating 2,250 kW of electricity. It was the largest electricity generation facility in Hong Kong. The electricity was used to drive its large machines and tools, and to light up shops and a dry dock (Middleton Smith, 1915). In

Deployment of Mond Producer

Brunner, Mond & Co., Ltd., Northwich
The Farnley Iron Co., Ltd., Leeds
Albright & Wilson, Ltd., Oldbury
Monks, Hall, and Co., Ltd., Warrington
The United Turkey Red Co., Glasgow
J. Brown and Co.
The Castner Kellner Alkali Co., Cheshire
J. and H. Robinson Ltd.
Wm. Beardmore and Co.
The Crossley Bros., Ltd., Manchester
Hollins Mill
Kosmoid Ltd.
The South Straffordshire Mond Gas Co., Tipton
The Salt Union, Ltd., Liverpool
The Trafford Power and Light Co., Ltd., Manchester
D.&W. Henderson & Co., Ltd., Glasgow
Cochrane & Co., Dudley
Ashmore, Benson, Pease & Co., Ltd., Stockton-on-tees
The Premier Gas Engine Co., Ltd., Nottingham
Handyside & Co., Ltd., Derby
The Railway and General Engineering Co., Ltd., Nottingham
Tweedales & Smalley, Castleton, Manchester
Cadbury Bros., Ltd., Birmingham
University of Birmingham
Blair and Co. works
Midland Railway Works, Heysham Harbor

Fig. 9.5 Industrial scale deployment of Mond gas producer (*Data source* Allen, 1908; Co, 1903)

Japan, two companies had already deployed the Mond gas producers in 1915, one at the Omori coal gasworks in Tokyo and the other with the Mitsui Mine Co., Ltd. at Ita, Fukuoka prefecture, which was used as prime driver. The Omori coal gasworks is one of the early coal gasworks developed by Tokyo gas probably in late 1880s to supply illuminating gases (Nakai, 1915).

With the Mond gas producers being recognized and becoming deployed by industries, steam engines started to feel the pinch from the competition with the clean and efficient gas engine, kicking off the age of gas engines of the twentieth century.

Mond Gas and Electrons

At the time when Mond began to tackle the issue of ammonia making, technology for generating electricity was evolving swiftly as well. After showcasing his incandescent light bulbs in 1879, Thomas Edison (1847–1931) moved on to build the first complete central station on Pearl Street in New York City and dispatched electricity to 1,400 incandescent lamps within a radius of a mile in September 1882. The complete central station includes steam generators, turbines, dynamos, voltage regulating devices, a switchboard, copper wire conduits laid underground and fixtures that were used to mount lamps at destinations. Each of the six dynamos weighed 27 tons designed with a capacity of 100 kW. For a gas engine application, a 100 kW dynamo would require a gas engine of about 167 HP to provide the needed driving power, which translates to a gas producer processing about 1.8 tons of coal daily. At this scale, gas engines using either coal gas or Dowson gas would be hardly competitive to steam engines. For the generation of electricity the trend had already been set that large scale and centralized stations would seem no doubt to be the future.

Once improved the incandescent light bulbs and adopted a three phase alternative current (AC) generator Edison incorporated under Edison General Electric in New York on April 24, 1889, by merging three of his manufacturing companies and the Edison Electric Light Company held the rights of all intellectual properties. The company speeded up its pace to develop and build more central stations with steam engines in America and beyond.

By 1900 many coal fired steam power plants including central stations between 1–10 MW had been developed. This development had pretty much put a fatal end to the coal gas for illumination in cities and then towns where central stations and grid networks for electricity distribution had been built. A shortcoming of the steam engine system for electricity generation, however, was its low efficiency. Some of the systems had an actual efficiency of as low as 4% (Fig. 9.6). Therefore, electricity was not cheap because generating 1 kWh of electricity would require 6 pounds of coal at such a performance. Taking a closer look at the steam engine system for electricity generation, the technology around 1900 was still in a primitive stage. The coal firing furnace, boiler, low steam pressure (about 180 psi), saturate steam or slightly superheated, and the cylinder type steam engine were essentially the same as decades ago. Steam engine to electricity cycle

Efficiency, %	Coal to steam/gas fuel	Steam/Gas fuel to electricity	Overall
Coal combustion	59.0	6.8	4.0
Coal gasification	84.0	19.7	16.5

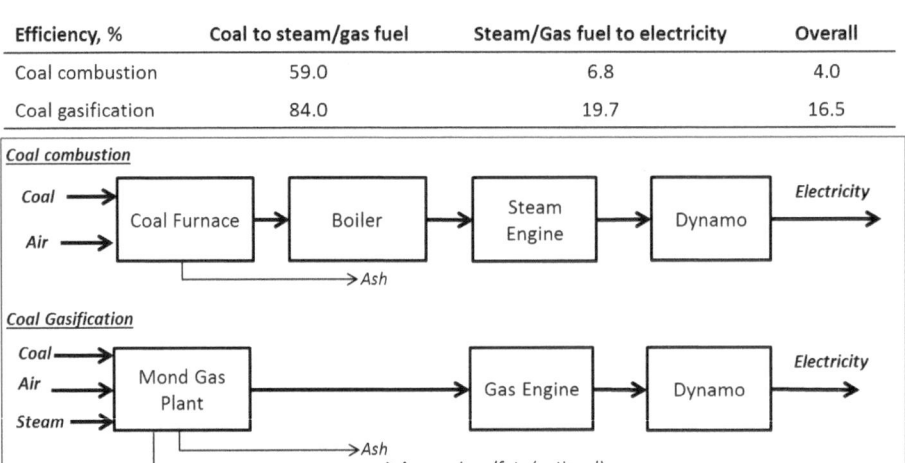

Fig. 9.6 An alternative generation of electricity from coal (Co, 1903)

performed at an efficiency of less than 7% because coal to steam cycle had less than 60% of heating value in the coal transferred into the steam. Although many academics predicted that the system at the steam pressure of 180 psi should theoretically be able to generate 1 kWh of electricity out of 3.28 pounds of coal, or 3.03 pounds if running continuously on a 24 h basis, which only correspond to efficiencies of 5.3% and 8.0%, respectively. Noting that some stations had to idle their generation equipment due to no or little demand during off-peak time steam boilers, nevertheless, had to stand by during the time of being idle. The stand-by contributes to an additional waste of coal. There seems no better way to handle the demand change at the time. The hot stand-by of a boiler would allow it to restart quickly when electricity demand returns. Otherwise, it would take much longer time to restart a boiler from a cold state, a waste of time. In addition, a coal fired system required good quality coal of 14,211 BTU/lbs., normally costing 12 s/ton against the bituminous slack of 7 s/ton, delivered to gate and used with the Mond gas producer. Remember, another important reason that steam boilers had to use good quality coals was to abate the emission of black smoke from its chimneys.

Mond appears confident that gas engines should be more efficient than steam engines for electricity generation. The success recorded with a 75 kW and 100 V system back in in 1897 actually showcased that coal gasification integrated with gas engines or the early version of an IGCC project delivered an overall efficiency of 16.5% for electricity generation. The gas cycle efficiency stood solid at 19.7%, almost three times that of a steam cycle, representing a significant advancement in technology (Fig. 9.6). There seems no doubt that businesses and industries would be eager to take the advantage of this cleaner and more efficient alternative for electricity generation.

The ground work laid by Humphrey at the Winnington plant during the time of 1890s proved to be effective and timely. In the pursuit and hardworking, of course, Humphrey had become well versed with the Mond gas process and the principle of coal gasification which probably helped him develop the future Humphrey pump. Beginning in the late 1890s, the Mond gas process became embraced as a powerful tool by industries, cities and towns, either to improve the bottom line of their businesses, to upgrade technologies, and to combat public nuisances by eliminating those polluting chimneys. Here are a few examples of electricity generation with the Mond gas.

Two miles southwest of the Winnington soda plant there is a Village of Hartford with 20,000 inhabitants. The village council through the Northwich Electric Supply Co. originally planned to build a hydro station to generate electricity for lighting. The Council in 1897, however, changed its plan to instead develop a central station near the soda plant by drawing the Mond gas via a 12″ pipeline over the fence to drive its gas engines coupled with dynamos for electricity generation. The station had three sets of gas engine dynamo, 60 kW each set, with a total capacity of 180 kW at 480 V. During off peak time, typically between 12 am and early morning time, the station used a battery of 200 amperes-hour to supply electricity, which was backed up by two small dynamos. The two-mile DC line delivered electricity to the homes and streets in the village. Hartford claimed to be the first village to adopt electric lighting in England. It is certainly quite a unique and sophisticated system at the time that seems still relevant in today's environment. Most interestingly, its electricity price was only a small fraction of that of all other public electric supply companies established across England by 1902. The Mond gas sold over the fence at 2d every 1,000 cubic feet.

For electricity generation with a gas engine, many factors would impact the overall efficiency such as how a gas engine is coupled with a dynamo generator, and if a producer or the system is fully loaded or operated at a partial load, and so on. A gas engine can be coupled with a dynamo directly or indirectly. A direct coupling is typically more efficient than an indirect coupling because the use of a belt to convey driving force to the dynamo would cause additional mechanical loss. In October 1904 Mond Power Gas Corporation installed its Mond gas process for the Blair and Co.'s engineering works at Stockton-On-Tees to find out the performances of the dynamos of different makers and different ways of coupling. Four new Premier gas engines were deployed in four sets of combination as follows. The direct current was generated for onsite consumption.

- Set 1: a 250 HP gas engine directly coupled with a 140 kW Scott and Mountain dynamo
- Set 2: a 250 HP gas engine directly coupled with a 140 kW Westinghouse dynamo
- Set 3: a 250 HP gas engine rope driving a 140 kW Scott & Mountain dynamo
- Set 4: ditto as set 3

The four sets would have a total capacity of 560 kW. Because of an existing 60 HP gas engine onsite already it seems that the Mond gas producer was designed to produce enough Mond gas to provide a total of 1,060 HP driving power. This could be achieved by having one Mond producer gasifying about 10–11 tons of coal daily or two 5–5.5 TPD producers to do the same job. While the latter case may be a bit more expensive, but it would offer more flexibility when a partial operation was frequently required. Upon completion of construction, the Mond Power Gas Corporation hired a third party to carry out the performance tests of 6 h duration, two tests on Set 1 and Set 2 in December 1904 and two more tests on all four sets in April the following year. As Professor William, Ph.D. and lecturer in Fuel and Metallurgy, the Victoria University of Manchester concluded in his report, dated May 31, 1905, to the Mond Power Gas Corporation, *"……In conclusion, I desire to say that in respect of the gas, cleanliness, and uniformity of composition of the gas, this plant left nothing to be desired. In all respects, also, it was working very satisfactorily* (Allen, 1908)". From a technology perspective, with the directly coupled gas engine and dynamo sets exhibited an overall efficiency of 16.7% against 14.0%, which were coupled with either a rope or a belt. Although the fact that the former test ran with a good quality coal (Auckland Park coal) and the later test with bituminous slack may complicate such a comparison the major cause for the difference in efficiency should stem from the ways of coupling method as the Mond producer is relatively less sensitive to coal quality.

In the meantime, Mond Power Gas Corporation continued its efforts to grow its engineering and technology capability by expanding its feedstock portfolio to include lignite, peat, woodchips and sawdust etc. More Mond gas had found its use at electric stations, iron and steel works, foundries, smelting works, glass works, chemical works, and coal gas works etc. To make it more efficient Mond increased its producer capacity to 30 TPD and even larger. The later expansion of the South Staffordshire Mond Gas Co.'s plant site at Tipton added four 30 TPD Mond gas producers. The subsequent deployment of the Mond gas producer was a quick one. By 1918, the Mond gas producers had gasified 2.75 mm tons of coals annually, equivalent to 376 Mond producers assuming each having a capacity of 20 tons of coal daily (Source: Grace's Guide to British Industrial History).

In 1958 the company formed a joint venture with the Humphrey & Glasgow and John Thompson to target business in the nuclear industry related to process and treatment. In 1960 Mond Power Gas Corporation merged with Davy-United to form Davy-Ashmore, which is part of Davy Johnson currently.

In addition, Mond and his assistants became the earliest chemists, if not the first one, to analyze the coal-N and to tackle its behavior during a gasification environment between 1879 and 1895. Of course, the fact that coal contains nitrogen and releases it partly as ammonia when subjected to heating in isolation from the atmosphere had been a well-known fact by 1879 and many of the 1,600+ existing coal gas works were actually recovering ammonia or ammonia sulfate as by-products from their operations. What Mond and his assistants did, by quantitatively analyzing the coal-N and the factors that impact its conversion to ammonia did fill a gap, as small as it may be, in advancing the fundamental understanding of coal gasification. Then, coal-N received little attention until when the United States enacted the Clean Air Act in 1970s to address the emissions of NOx from coal fired facilities.

Looking back, what Mond had achieved in such a conspicuous way by building one industrial empire after another, the Solvay soda process taking over the sector in twenty years, the high quality nickel refining business in about fifteen years, the Mond gas business that took almost his life career and many more. What is interesting is that each of the businesses all started from a scratch or with just an idea. His approach to bringing those ideas into industrial products by carrying out extensive laboratory searches and trials to develop the necessary specific knowledge and engineering know-how to generate the intangible products for industrial scale demonstrations, and then by going back and forth between laboratory and field demonstration to refining the knowledge and know-how to apply back to the products, proves to be an effective model for innovation and technology development. Just like what Mond presented in his speech to the Society of the Chemical Industry in 1889 that "*the steady, methodical investigation of natural phenomena is the father of industrial progress*" exemplifies the importance of R&D to the development of an industrial process technology, setting up a model that has been followed ever since by companies and corporations. It also became a tipping point between the interactions between science, chemistry, and industrial process development. Prior to this point, engineers and inventors had been driving the technology innovation while chemists was trying to catch up and to develop a kind of knowledge or a theory as an explanations on what was going on.

Applied chemistry came to being.

References

Abel, C. (1884). *Patent No. 1464.*

Accum, F. (1819). *Description of the process for manufacturing coal gas.*

Allen, H. (1908). *Modern power gas producer practice and applications.* D. van Nostrand Co.

Co., R. W. (1903). *Mond gas.* R.D. Wood & Co.

Donnan, F. G. (1939). *Ludwig Mond, F.R.S.: 1839–1909.* Institute of Chemistry of Great Britain and Ireland.

Humphrey, H. F. (1897). The Mond gas-producer plant and its application. In *Minutes of the Proceedings of the Institution of Civil Engineers* (pp. 129, 190–217). Inst C.I.

Keen, F. N. (1901). *Memorandum on the Mond gas scheme.* Walter King, II, Bolt Court.

Langer, L. M. (1888). A new form of gas battery. *Proceedings of the Royal Society*, 296.

Larter, J. E. (1920). *Producer gas.* Longman, Green and Co.

Londoner, A. (1912). Municipal gas lighting in England. *The American Gas Light Journal*, 321–323.

Lunge, G. (1884). The sulfuric acid and alkali trade of England. *Journal of Chemical Industry*, 470.

Middleton Smith, C. A. (1915). Electric generating station in China. In *Proceedings of the Institution of Electrical Engineers* (p. 162). E. and F. N Spon.

Mond, L. (1886). *Patent No. 65.*

Mond, L. (1889). President's address onproduction of ammonia from coal. *Journal of the Society of Chemical Industry*, 505–510.

Nakai, S. (1915, July). On Mond gas for motive powers. *The Journal of Chemical Industry, XVIII*, 209.

Rambush, N. E. (1923). *Modern gas producers.* Benn Brothers Ltd.

Readman, J. (1883). *Patent No. 5359.*

Samuel Clegg, J. (1841). *A practical treatise on manufacture and distribution of coal-gas.*

Townsend, C. A. (2003). *Chemicals from coal, a history of Beckton products works.* Retrieved September 2022, from Greater London Industrial Archaeology Society: http://www.glias.org.uk/Chemicals_from_Coal/

Travis, A. S. (2018). *Capture nitrogen—The growth of an international industry (1900–1940).* Springer.

Wisniak, J. (2006, January). *Ludwig Mond—A brilliant chemical engineer.* Retrieved 2022, from ResearchGate: https://www.researchgate.net/publication/236232462

Chemistry and Industrial Synthesis

10

By 1900, producer gas had been widely deployed in almost every corner of the industrial sectors in England from electricity generation, mechanical power, boiler heating, fueling kilns and furnaces, large and small, to iron and steel making. There were about 150 engineering companies in both Europe and the United States involved in manufacturing and building producer facilities. The demands for large gas engines by the iron and steel industry and the centralized electricity generation had created the urgency for large gas producers. In Germany, the precipitous growth of its iron and steel industry continued its momentum that had created a significant demand for large gas engines to provide prime drivers for its large blowers of 2,000 HP or larger at steel mills.

What happened in the United States was similar but differed in terms of the technology adopted. Lowe's water gas technology continued to dominate the manufactured gas market. When Lowe's key patents became expired around 1892, more companies, contractors and fabricators got involved in marketing and developing water gas projects. These companies include the Gas Machinery Company (Cleveland, Ohio), the Bartlett-Hayward Company (Baltimore), the Gas Engineering Co. (Trenton, New Jersey), the Western Gas Construction Co. (Ft. Wayne, IN), the Koppers Co. (Pittsburgh), and the West Gas Improvement Co., of Britain. It started to penetrate the markets of England, the European continent, and the rest of the world. The carburetted blue water gas had reached its peak about 1926, representing 58% of all manufactured gases in the US market (Morgan, 1945). In the meantime, water gas process had evolved into versatile forms that fit the needs of project requirements by different owners. So did to the gas producers. In 1921, there were about 11,000 commercial gasifiers in US consuming more than 40,000 tons of coal daily. During next decade, however, the advent of NG and refined Petro products led

© The Author(s), under exclusive license to Springer Nature Switzerland AG 2024   143
Q. Zhuang, *From Coal to Hydrogen*, Synthesis Lectures on Chemical Engineering and Biochemical Engineering, https://doi.org/10.1007/978-3-031-55586-2_10

to a rapid decline of coal gasification deployment in US. By 1948, there were only about 2,000 gasifiers in service (Corp, 1980).

Entering 1910, the advancing of chemistry especially in the field of synthetic chemistry in Germany had created a few consequential fundamental changes, not only in the field of coal gasification but also in the development of human society. The first comes the commercialization of an artificial or synthetic ammonia project that was put into operation by BASF in 1913, followed by another industrial production of methanol from synthetic gas or syngas, primarily hydrogen and carbonic oxide, in 1923, and the coal liquefaction plants developed from the mid-1920s, opening up a new age for coal gas. Such an industrial development opened up a huge market for coal gas making or coal gasification, either gas producer or water gas generator, which the former typically provides process heating while the latter produces hydrogen and carbonic oxide as raw materials going to the synthetic plants. To meet such demands, however, gasification had to overcome a few challenges that hinged on some fundamental developments not only of its own but also of a few downstream technologies including gas separation and shifting because each of the individual synthetic processes requires its own set of requirements for syngas. These in turn call for clarity of some fundamental questions about what matters are and how matters interact with each other chemically.

Then, the molecular theory reemerged.

Molecular Theory

Ever since the Siemens brothers put the gas producer into commercial use in 1861 Engineers and chemists had been looking for ways to explain what is happening inside a gas producer and the factors impacting the overall performance of the producer. Such a need had become pressing when Dowson tried to establish his compact producer package to deliver clean fuel gas to service gas engines and when Mond spent most of his career developing a large scale process that is capable of processing low quality bituminous coals to produce a clean and cheap fuel gas for a large industrial adoption. To better design a producer and to make it run more efficiently, the need to know what the gases are fundamentally had become obvious without which it would be impossible to explain what happens inside a producer. Due to the lacks of knowledge, however, engineers and chemists at the time had so far been vague and general when it comes to the fundamentals about what the gases are and what happens inside a producer. A few examples are that the single element gases such as oxygen, hydrogen, and nitrogen had been generalized as O, H, and N. The atoms as Dalton defined do not necessarily determine the property or characteristics of a matter or a substance, which is actually determined by how the reacting atoms are combined. Moreover, Dalton's atomic theory could not provide the fundamentals that govern how those atoms would combine and how the surrounding conditions would impact such combinations. In a producer environment, when a carbon reacts (combines) with an oxygen they always form either carbonic oxide or carbonic acid but not any other random forms. Chemists started to realize that Dalton's atomic theory faced more challenges than dilemmas.

While Dalton's gravimetric approach focused on the weight of atoms, his French counterpart like Gay-Lussac took a volumetric approach in his experiments by measuring the volume of gases to interpret the nature of the gases and their interactions. This led to the establishment of one of his most significant theories, known as the law of the combining volumes of gases. Gay-Lussac found that when a molecule of hydrogen and a molecule of oxygen combine to give a molecule of water the two gases always follow a volumetric ratio of 2–1; the combination of a hydrogen chloride with ammonia always goes by an equal volume to form an ammonia chloride. Based on these observations a few years later in 1811 the Italian Physicist, Lorenzo Avogadro (1776–1856), proposed a concept of molecules as the smallest particles, instead of atoms, that make up matters. It is the molecule that determines the property of a matter or a substance. Avogadro's concept calls that an equal volume of gases at the same conditions of temperature and pressure must contain the same number of molecules. Based on the fact that one volume of hydrogen and one volume of chlorine give two volumes of hydrogen chloride Avogadro reasoned that there must be twice as many molecules in either one volume of hydrogen or one volume of chlorine as in one volume of hydrogen chloride. Then, the logic leads to that there must be two particles in each of the original hydrogen gas and chlorine gas; they are not indivisible atoms as each particle of hydrogen chloride contains both hydrogen and chlorine. This is a significant breakaway from the 'common knowledge', Dalton's atomic theory, at the time. Following Avogadro's molecular theory it would be a step closer to explain what happened inside a producer. That is, the different combinations when a carbon reacts with an oxygen molecule, the reaction would result in two completely different molecules, carbonic oxide or carbonic acid, which the former is inflammable but the latter not at all but rather suffocating a flame. When a carbon reacts with a water molecule, a flame distinguishing agent would generate combustible gases of hydrogen and carbonic oxide. Although it would take a few more steps to understand the theory of chemical bonding this would put engineers and chemists on the right track when it comes to interpreting the chemistry or reactions that take place inside a producer and a water gas generator. Avogadro's molecular concept, unfortunately, had been ignored for about a half century. While there must be some reasons to blame, the fast development of electrochemistry during the first half of the nineteenth century appears one of the significant at the time. Plus, the Swedish Chemist and Physician Jöns Jacob Berzelius (1779–1848) is one of the most influential individuals who was right in the center of the established electrochemistry.

Berzelius, one of the strong proponents of the Lavoisier new chemistry framework, is one of the major founders of modern chemistry for his contributions such as to Dalton's atomic theory and chemical symbols, and for his development of the electrochemical theory etc. Since 1800 the voltaic pile that Volta invented had become a powerful tool for

scientific exploitations in different fields. In addition to its applications in the electro-
magnetic field Davy Humphrey in about 1803 applied the voltaic pile to isolate elements
of potassium and sodium from their respective molten salts. Similarly, Berzelius adopted
an electrolytic approach extensively to investigate a wide range of inorganic substances
between 1803 and 1844. His publications such as the 1819 *"Essai sur la théorie des
proportions chimiques et sur l'influence chimique de l'électricité (Essay on the Theory
of Chemical Proportions and on the Chemical Influence of Electricity)"*, and the 1826
"Lärbok i kemien (Textbook of Chemistry)" had been widely read. Such an extensive
investigation allowed Berzelius to generalize his findings by formulating the principle
of an electric dualism that forms part of modern chemistry. Based on the principle, any
substances, either naturally occurring or artificially made, are all comprised of two electrically opposite components. For example, water, sodium chloride, sodium carbonate and
Calcium sulfide are all made of an electrically positive cations of H^+, Na^+, Na^+, Ca^{2+}
and electrically negative anion of O^{2-}, CO_3^{2-}, Cl^- and S^{2-}, respectively. Based on the
principle, Berzelius implemented the acid/base theory, a step forward from what was originally proposed by Lavoisier. Berzelius further extended the electric dualism principle to
that only electrically opposite elements or atoms could combine to form a compound and
it is impossible for any two identically charged elements or atoms to combine to form
any compounds. At that time, this school of thought seems very logical and perfect in
line with the understanding and knowledge of electricity and its subsequent effects on
cohesion and expulsion of Newton's mechanical chemistry. It is, however, problematic
for those engineers and chemists working in the field of coal gasification simply because
many gases involved in a gas producer or a water gas generator such as oxygen, hydrogen
and nitrogen would only exist in a single atom or an element as O, H, and N. In another
way, it is impossible to have a diatomic structure. Actually, most of the publications and
writings associated with coal gas making including those by Sir William, Dowson, and
Mond etc. presented these gases (oxygen, hydrogen and nitrogen) as a single atom element for most of the nineteenth century. In addition, carbonic oxide and carbonic acid
had often been expressed as C_2O and CO, respectively or, in many occasions, simply
left the formula out to prevent confusion. Such confusion continued until the emergence
of organic chemistry that started to create some serious questions and challenges to the
legality of the atomic theory and the principle of the electric dualism or the ionic theory.

From about 1820s when the by-product coal tars produced from the fast growing coal
gasworks had become the target of many workouts, studies and investigations aimed at
finding ways either to dispose of it or to utilize it. Such developments facilitated the forming of organic chemistry because the organic compounds discovered from the coal tars
tend to be much more concentrated, complex and characteristic compared to others found
in vegetables. Since Frederich Woehler (1800–1882), a German chemist who acquired his
chemical analytical skills from Berzelius while in Stockholm, shared his first synthesized
urea in early 1928 chemists had identified by 1850 many organic compounds from coal tar
such as BTX, naphthalene, anthracene, aniline, quinolone, pyridine, and phenol etc. These

compounds or substances are completely different from inorganic compounds. These compounds are from completely different origins, coal tars, living bodies and plants, which are primarily built up by carbon and hydrogen. Both Dalton's atomic theory and Berzelius' electric dualism could hardly explain or be applied to these compounds. The status quo, however, continued for years most probably due to the fact that no commercial values of these compounds had been developed at that time, which might have not generated the deserved urgency for a fundamental understanding. When William Perkin (1838–1907) "accidentally" discovered the first synthetic dye, the Mauve, in 1856, coal tar suddenly became the black gold overnight, which led to the establishment of the most valuable chemical industry, the dyestuffs. Such a commercial development might have created some urgency to the scientific world. In 1858, Stanislao Cannizzaro (1826–1910), an Italian chemist, published his essay *"Sunto di un corso di filosofia chimica (Summary of a Course in Chemical Philosophy)"* in *Nuovo Cimento (New Trial)*, which was reprinted as a pamphlet (From Alchemy to Chemistry, 2000). In it, Cannizzaro basically presented his interpretation of the Avogadro's molecular theory about atoms, molecules, and the determination of their weights. Then in 1860, Cannizzaro made the pamphlets available at the Conference of Chemistry at Karlsruhe on Sept 3–5, 1860, which he believed it would help clarify the confusions about Dalton's atomic theory and Berzelius' principle of electric dualism. In retrospect, the Karlsruhe conference is historical not only because it was the first ever international conference of chemistry but because it was designed to resolve the pressing issue on atoms, molecules, and the structure of the periodic table with which chemists had been struggling for quite a while. Cannizzaro's lecture along with the small pamphlets distributed seems to ring the bell in the long run that Avogadro's molecular theory would make more sense than either Dalton's atomic theory or Berzelius' electric dualism principle in the face of many questions around coal gas and coal tars. From the point on, the molecular theory started to weigh in more and more and eventually took its shape. When the Dutch physical chemist Jacobus H. van't Hoff (1852–1911) made another pamphlet public about his theory of valence bonding between a carbon and a hydrogen around 1873 the molecular theory began to land on a firm ground (per Britannica). In essence, it is Avogadro's molecular theory, instead of Berzelius' electric dualism, that should be applicable to organic compounds and gases including those involved in coal gasification. Gases like oxygen, hydrogen and nitrogen are diatomic in nature and exist in the form of O_2, H_2 and N_2. Carbonic oxide should be mathematically expressed as CO and carbonic acid as CO_2. Although the debate over the nature of chemical bonding continued into the twentieth century the actual adoption of molecular theory had provided enough foundation for gasification to move a step further to paint a clearer picture on what is going on inside a gas producer, a water gas generator or a gasifier in general. But, it took some time to discover.

In 1870, William Crookes and Ernst Roehrig started to express carbonic oxide and carbonic acid as CO and CO_2 in their book *"Practical Treatise on Metallurgy"* but left the water as HO and the diatomic molecules as X (O and N) (Roehrig, 1870). Then, the efforts of standardization on chemical nomenclature made by the Royal Institute of

Chemistry in 1882 and by ACS in 1884 had certainly made the mathematical and sto-
ichiometric expression of gasification reactions much easier, if not impossible. Around
the same time, the development of thermodynamics led to the establishment of the low
temperature liquefaction and distillation technique, the cryogenic technology, which as a
powerful tool made it possible to isolate many of the individual gases with a high purity
so that chemists and physicists could investigate the physical and physicochemical properties of each of the individual gases such as O_2, H_2, N_2 and more, certainly contributing
to the establishment of the molecular theory.

In 1905, the American Engineer Samuel Wyer started his book "*Treatise of Producer-
Gas and Gas-Producer*" by putting together a summary of all fundamental laws
that govern gases, their definitions and methodology to determine the physical and
physicochemical properties of the gases available at the time. Following are a few
examples.

§ 14. Joule's law of gases.

No change of temperature occurs when a perfect gas is allowed to expend without doing
external work, or without taking in or giving out heat.

§ 15. Law of Gay-Lussac.

Equal volumes of all gases at the same temperature and pressure contain the same number of
molecules.

§ 16. Dalton's law.

A mixture of gases, having no chemical action on each other, exerts a pressure which is equal
to the sum of the pressures which would be produced by each gas separately, provided it
occupied the containing vessel alone at the given temperature.

......

§ 31. Atoms and molecules.

An atom is the smallest particle of an element that can enter into chemical combination.
Atoms combine to form molecules of substances.

For comparison, the following definition of Joule's laws should be more relevant and can
be found in most modern dictionaries. "*(1) the principle that the rate of production of
heat by a constant direct current is directly proportional to the resistance of the circuit
and to the square of the current; (2) the principle that the internal energy of a given mass
of an ideal gas is solely a function of its temperature.*" The *principle (2)* is what governs
the expansion of an ideal gas without doing external work, an adiabatic expansion, during
which there would be a drop in the temperature of the ideal gas, which forms the basis
of the cryogenic technology. The molecular theory seems to have been accepted except
for the fact that it is not the atom but an electron and a proton the smallest particles,
which became known in the early twentieth century. When describing the reactions that
take place inside a producer or a water gas generator, however, Wyer was still used to the
following.

$$C + 2O = CO_2$$

$$CO_2 + C = 2CO$$

$$H_2O + C = 2H + CO$$

Obviously, Wyer either did not pay attention to the development of the molecular theory or did not feel the value of applying it or still did not believe the fact that a diatom would exist. It more often than not takes an extra time for the diatomic nature of the molecules of oxygen and hydrogen to be accepted.

This is about to change; a year later in 1906, a peculiar mind who had been active in the gas producer business put the up to date molecular theory into use.

Fundamentals of Coal Gasification

At the beginning of the 1900s, most of the coal gas making including gas producers had been focusing on experience and hands-on expertise. The experimental investigations, for example carried out by Mond and his assistants in laboratory and field scale facility, focused on producing the rules of thumb and specific knowledge used to design, build and operate a gas producer. Dowson appears a very peculiar individual. In addition to running his gas producer business Dowson had spent quite some of his time and resources looking into the fundamental part of what happens inside a gas producer. Dowson started his book published in September 1906, *"Producer Gas"*, by addressing the need for a theoretical understanding (Larter, 1907).

> Producer gas has now a recognised position in practical work, and I am confident that in the near future its adoption will be greatly extended; it is, therefore, desirable that its production and application should be considered theoretically as well as practically. Mere rule of thumb cannot be right. I trust, therefore, that this little work may be the forerunner of some more complete and exhaustive treatment of the subject. There are numerous books on the theory and practice of the gas-engine, but so far as I am aware there is no complete work on producer gas.

Dowson must have invested resources to investigate the fundamental aspects of the chemical reactions taking place between coal and air/steam inside the environment of a gas producer. Because of his extensive experience with coke or anthracite coal, Dowson was able to simplify what happens inside a gas producer into a chemical reaction system among carbon, oxygen and steam. Then by applying two principles established thus far, Dowson was able to estimate or predict the temperature and the composition of producer gas based on a set of assumed parameters. The first principle he applied is the mass balance around a gas producer operation, that is, what's in equals what's out, which was leveraged by Lavoisier more than a century back. Then he also applied the rule of balance to energy around a gas producer, the first law of thermodynamics demonstrated

Table 10.1 Major chemical reactions in a gas producer

Reaction		Enthalpy (kcal/kmol)	Remark
1	$C + O_2 \rightarrow CO_2$	−97,600	Combustion
2	$C + CO_2 \rightarrow 2CO$	+38,800	Gasification
3	$2C + O_2 \rightarrow 2CO$	−58,800	Reaction (1) + (2)
4	$C + H_2O \leftrightarrow CO + H_2$	+28,800	Water gas making
5	$C + 2H_2O \leftrightarrow CO_2 + 2H_2$	+18,800	Ditto
6	$CO + H_2O \leftrightarrow CO_2 + H_2$	−10,000	Water gas shifting
7	$C + 2H_2 \leftrightarrow CH_4$	+92,380	Rambush (1923)

by Joule more than 60 years ago. By 1900 physicists and physical chemists had established adequate data on most individual gases thanks to the isolation of the pure gases made possible with the cryogenic technology, which is based on the second law of thermodynamics. In the meanwhile, as Dowson pointed out that gaseous reaction had been extensively investigated both fundamentally and practically in a gas engine environment, which provided the needed information such as combustion heat between gases as well. The available information could make a detailed thermodynamic analysis of a gas producer possible. Actually, for example, the well-known Boudouard reaction that carbon monoxide disproportionates to carbon dioxide and carbon was discovered and extensively studied around 1901. With the development of the molecular theory, Dowson was able to put the package of available information into work and painted a picture of what happens inside a gas producer mathematically, becoming probably the earliest attempt at the time (Table 10.1).

Coal gas making is simply a thermal process, coal decomposing or dissociating upon attack by heat or oxidants such as oxygen, steam and carbon dioxide etc., under the required conditions. By simplifying a gas producer into a carbon–oxygen-steam system the reaction zones shown in Table 5.1 would essentially become two zones of combustion and gasification, instead of three or four, because there would be no volatile matter and moisture involved. To start with his modeling Dowson in his first step dealt with only a carbon–oxygen system, which made the model much simpler.

Here is how Dowson described mathematically what happens inside the gas producer. Oxygen blown in from beneath the grate reacts quickly to exhaustion with carbon in the zone of combustion, which is typically a very thin layer, right above the grate following **Reaction (1)** forming carbon dioxide, which moves up into the zone of gasification and completely reduced by carbon to carbon monoxide following **Reaction (2)**. To address a debate at the time about if carbon dioxide is the only product formed in the zone of combustion Dowson also believed that **Reaction (3)** would take place as well. In the meanwhile, **Reaction (3)** represents the sum of **Reaction (1)** plus **Reaction (2)** or the overall model reactions between carbon and oxygen taking place in the gas producer;

two k-moles of carbon react with one k-mole of oxygen gives off two k-moles of carbon monoxide, which mass is balanced. This is a close approximation at high temperatures. Now Dowson applied the first law of thermodynamics by stating that **Reaction (1)** releases 97,600 kcal of heat where 38,800 kcal is absorbed by **Reaction (2)**, ending up with 58,800 kcal of heat released into the environment inside the gas producer. The total energy stored with two k-moles of carbon following combustion **Reaction (1)** would be $2 \times 97,600$ kcal. Therefore, the energy retained in the producer gas, the two k-moles of carbon monoxide, would be $(2 \times 97,600$ kcal $- 58,800$ kcal$)$, resulting in a cold gas efficiency of $(2 \times 97,600$ kcal $- 58,800$ kcal$)/(2 \times 97,600$ kcal$) = 70\%$. Of the released heat inside the producer, typically 22% becomes a sensible heat of the hot producer gas and about 8% is lost through radiation and conduction to the surrounding of the producer. To predict how high the producer temperature could achieve Dowson included air back into the scenario to replace oxygen and the **Reaction (3)** now becomes the following.

Model 1: $2C + O_2 + 3.76N_2 \rightarrow 2CO + 3.76N_2$

Considering air containing 21% oxygen and the rest as nitrogen there is 3.76 k-mole of nitrogen for every one k-mole of oxygen. Because nitrogen in the air would not impact heat balance or chemical reactions, by assuming the heat capacity of carbon monoxide and nitrogen are of the same (0.245 kcal/kg/°C), the available sensible heat to raise the producer gas to 1080 °C. The producer gas would have 34.7% of carbon monoxide and 62.7% of nitrogen under the model assumption (Table 10.2), which would have a heating value of 1,060 kcal/ m^3 or 118 BTU/cf. Dowson's work had become the earliest theoretical estimation of a gas producer's performance.

Then Dowson moved on to introduce steam into the carbon and air reaction system. At a high temperature inside a gas producer steam would react with hot carbon to form carbon monoxide and hydrogen known as the water gas via **Reaction (4)**, which is typically the primary reaction. In the meantime, **Reaction (5)** to form carbon dioxide and hydrogen takes place as well to some extent depending on the producer conditions, which is further moderated by **Reaction (6)**, the water gas shift reaction. Due to the endothermic nature of the **Reactions (4)** and **(5)** the introduction of steam into a gas producer would help moderate, reduce and smooth out potential spikes in gasification temperatures inside

Table 10.2 Gas compositions of Dowson's modeling

Gas composition	Model 1		Model 2	
	kmol	Vol%	kmol	Vol%
CO	2	34.7	3.48	39.9
N$_2$	3.76	65.3	3.76	43.1
H$_2$	–		1.48	17.0
Subtotal	5.76	100	8.72	100

the gas producer, which is an important function that steam plays. Of importance is the **Reaction (4)**, which should always be maximized because it could boost a producer gas composition; a carbon reacting with a water gives off a carbon monoxide and a hydrogen, a thermal way to split water to produce hydrogen and carbon monoxide. This is what Sir William observed the 12.5% increase of the effective fuel gas once passing through the regenerator at a higher temperature. The gaseous phase **Reaction (6)**, however, is an exothermic and equilibrium reaction, which is favored toward the right direction, carbon monoxide to carbon dioxide, at a low temperature. This explains the relatively low concentration of carbon monoxide of the Mond gas when a low temperature is adopted to maximize the recovery of ammonia.

To be able to predict the gas composition of the carbon-air–steam model reaction system Dowson made further assumptions such as that no heat loss around the producer and the heat released from **Reaction (3)** would be used to decompose steam via **Reaction (4)**. He estimated that every kmol of carbon reaction in **Reaction (4)** would decompose about 0.74 kmol of steam. Then the model reaction 1 becomes the following.

Model 2: $2C + O_2 + 1.48H_2O + 3.76N_2 \rightarrow 3.48CO + 1.48H_2 + 3.76N_2$

There is no doubt that the *Model 2* is an over simplification of a carbon-air–steam system in a gas producer environment. In reality, of course, heat is not balanced and more carbon would be burnt to carbon dioxide to provide the needed heat; more air blast would, therefore, be necessary. This is why a typical producer gas would contain a certain amount of carbon dioxide and higher nitrogen than *Model 2* estimated. The significance of Dowson's pioneering modeling work under certain practical assumptions, however, is that such a theoretical analysis would provide guidance to the engineering and operation of a gas producer toward the desired products. Should carbon monoxide, for example, be the desired product, an operating temperature as practically high as possible and a short residence time of the resulting gases should be considered in designing and operating the producer because a high temperature would favor the **Reaction (3)**, **Reaction (4)** and **Reaction (6)**, which **Reaction (6)** would be further at a disadvantage to reach an equilibrium at a short residence time. The opposite holds true should hydrogen be desired. Choosing carbon as a model material allows Dowson to simplify the complex reactions taking place in an environment of a gas producer and to dissect each of the major reactions to analyze their characteristics and impacting factors, which was enabled by the established molecular theory; it is the chemical bonding between the reacting elements, either breaking up or uniting by a chemical reaction, that involves energy by releasing it or absorbing it in a chemical reaction(s) in order for a reaction to proceed.

In an actual environment using coal as feedstock, the reactions involved in a gas producer would become much more complex not only because of the large amount of existing volatile matter but also due to the impurities typically in a small amount such as nitrogen, sulfur, chloride and ash (inorganic salts or oxides). With regard to the impurities in an oxygen deficient or reducing environment inside a producer or gasifier majority of the coal-N in coal generally converts to gaseous nitrogen (N_2), with some ammonia (NH_3), and a small amount forming hydrogen cyanide (HCN). Most of the sulfur becomes hydrogen sulfide (H_2S) and a small amount forms carbonyl sulfide (COS). Chloride is primarily converted to hydrogen chloride (HCl). In general, the quantities of sulfur, nitrogen, and chloride in the fuel are sufficiently small that they have a negligible effect on the main syngas components of H_2 and CO. Dr. Rambush, chief engineer of *Mond Power Gas Corp*, focused more in his book *"Modern Gas Producers"* of 1923 on the impacts of contents of moisture, volatile matter and ash contained in the coals to the operation, performance and producer design. He sufficed his discussions with a significant amount of information acquired from his own experiments and those conducted by others during the early twentieth century. Of interest is the fact that Dr. Rambush pointed out the importance of understanding the quantity and composition of coal ash and its clinkering and slagging tendency in engineering and operating a producer. A methodology was provided to analyze the ash melting or deformation process, typically between 1100 and 1700 °C, which has been utilized in modern times. Dr. Rambush also explained the eutectic nature of coal ash and ways to change its melting behavior by adding different fluxants. From a theoretical perspective, Dr. Rambush also brought up the reaction of hydrogen with hot carbon to form methane in a gas producer environment, **Reaction (7)**, which is a strongly endothermic reaction along with several other similar reactions. It was explored later by Lurgi AG when developing its pressurized Lurgi gasifier to maximize the methane formation in order to produce a town gas of a high heating value.

In summary, the establishment of molecular theory provided a pathway to take a closer look at the reactions and chemistry taking place inside a gas producer or water gas generator by dissecting and analyzing each individual reaction. Such analysis proves beneficial as guidelines for the operation of a gas producer or a water gas generator and for their improvements as well. Following Dowson's pioneering work on theoretical modeling many efforts have been made afterward to mathematically express and predict the performance of a gas producer or water gas generator. As Dr. Gumz pointed out in his 1950 book *"Gas Producers and Blast Furnaces—Theory and Methods of Calculation"* that *"A process is completely mastered only if it can be calculated."* By applying the kinetic molecular theory that was later established, Dr. Gumz made additional mathematical analysis of chemical reactions inside a gas producer and blast furnace. Coal gasification in general remains to some extent the art of science and technology and requires hands on experience. The journey to master it still lies ahead.

Water Gas and Industrial Synthesis

After Mond's "attempt" to fix air-N to make ammonia around mid-1880, the dream of making ammonia out of air appears still strong. With the dust settled concerning the molecular theory, chemists had turned their focus on the reactions of gases between nitrogen and hydrogen, hydrogen and carbon monoxide, and even hydrogen and coal etc. in a belief that these reactions would take place should a right set of conditions be created; a condition extreme enough to break up what bond the two atoms together such as the $N{\equiv}N$, $H{=}H$ and $O{=}O$. A branch of chemistry emerging at the time made it possible; it is catalytic chemistry following the establishment of thermodynamics and physical chemistry. In the meanwhile, the establishment of the large scale industrial productions of coal gas, water gas and producer gas, had certainly laid a solid foundation in terms of the supply of hydrogen and carbon monoxide. In addition, the ongoing thirst for more fertilizers and the emerging demands for motor fuels also made the subject around catalytic synthesis more attractive. And this time, Germany had become the focal point for this development thanks to its long time commitment to the experimental scientific approach that led to its establishment of organic chemistry that had ultimately made Germany the world monopoly of dyestuffs, the products made out of coal tars.

Although William Perkin first developed the purple dye, the Mauve, in the 1850s, it is Perkin's mentor, August Hoffmann (1818–1892) whose pioneering work on coal tar led to the establishment of the artificial dye synthesis in Germany to a whole different level. By 1878, Germany had already made 63% in global sales of aniline, the critical feedstock to make dye and alizarin dye (Page, 1879). With more dyestuffs made, Germany soon became the dominant source of dyestuff supply by the turn of the century. Such a success, of course, was made possible by the easily available coal tars produced from both coal gas works and the fast growing iron and steel industry at the time. Iron and steel making requires a large amount of metallurgical coke, different from that of coal gas making, while the coal tar and coke oven gas ("COG") as by-products are left to be exploited. Then, entering into the twentieth century and, especially going through WWI the deployment of automobiles and by-planes made many believe that mechanization would be the spot that held a bright future. The state of Germany and many companies, therefore, began to look beyond coal tar and COG for next growth opportunity, the liquid fuels.

In Germany, the Lack of crude oil resources made its plentiful local coal supply the natural choice of feedstock going into the synthetic industry, a totally different field that requires a completely new while challenging set of conditions in order to make it a reality. Thereof, fixing air-N with hydrogen to produce ammonia sets off the journey of a synthetic industry that has ever since been widely practiced in different regions and countries where lack convenient resources such as crude oil and natural gas even in modern times.

Ammonia Synthesis

Fritz Haber (1868–1934) received his chemical training under Bunsen at the University of Heidelberg, under Hoffmann at the University of Berlin and under Liebermann at the technical school at Charlottenburg between 1886 and 1891. After some time of soul searching between an industry and an academic world Haber was somehow drawn into the field of combustion chemistry involving gases and hydrocarbons by becoming an assistant to Professor Hans Bunte, the chair for chemical technology at the University of Karlsruhe. Haber set his eyes on the research subjects that would have industrial applications. Among many works and inventions Haber studied the energy efficiency of steam engines, turbines, and motors, and ways to improve the efficiency by cutting energy loss in those systems. He also investigated the Bunsen burner and found out that the Bunsen burner flame has an inner cone and an outer cone where in the former a luminous water–gas equilibrium is established and in the later, there is the combustion of the water–gas. This is why the Bunsen burner is more efficient than the Argand burner used during the early years of coal gas lighting. It appears that Haber's interest in fixing air-N to make ammonia came when he visited a cyanamide based process plant built in Niagara Fall of the United States in 1902, where the available hydropower made the electricity cheap enough (Technology, 2019). Different from the attempts made by Mond and his colleagues 20 years ago, Haber understood that fixing air-N with hydrogen chemically would require an extreme temperature and pressure to break up the bondings of $N \equiv N$ and $H=H$ even aided with catalysts, which is a critical path to the ammonia formation. After being appointed Professor of Physical Chemistry and Electrochemistry and Director of the Institute established at Karlsruhe in 1906, Haber started his research work and soon made a significant breakthrough on a laboratory scale. Realizing the promising work of Haber's, BASF in 1908 took an interest in Haber's laboratory work on the technology to make artificial ammonia and bought the technology rights to jointly develop the necessary process and catalyst for an industrial scale production of ammonia. The joint development was under the direction of Carl Bosch (1874–1940), a chemist of BASF. Bosch also received technical training in metallurgy. About five years later on September 9, 1913, the first industrial ammonia plant at Oppau, about 3 km north of the BASF's Ludwigshafen complex, went on stream and produced the first ammonia. The Oppau plant was designed with a capacity of 30 tons of ammonia daily. The reaction of hydrogen with nitrogen was aided with an iron metal promoted with alkaline at 150–200 bars and 500 °C. There, the Haber–Bosch process was born, a significant milestone to finally fix air-N to make ammonia at a pressure unprecedented understanding that Mond producers were only operated at a pressure of a fraction of an atmosphere. Then, the Haber–Bosch process soon became a global brand and has enjoyed a rapid growth after the first production at the Oppau. Annual ammonia production on a nitrogen basis increased from about 4,000 tons in 1914, to 100,000 tons in 1920, and about 900,000 tons in 1930. By 1945, there were about 125 plants in operation with a capacity of over 4.5 million tons of ammonia (estimated in nitrogen) per year (Technology, 2019).

In fact, BASF's interest in ammonia production goes back to an earlier time when Bosch joined the company in 1899; he was assigned to investigate fixing air-N to produce ammonia via metal cyanides and nitrides. Such an effort had evolved into a pilot scale by 1907. The company obviously changed its direction once witnessed the potential of the technology that Haber had put together. It appears at that time, however, that BASF's interest was beyond ammonia. As a company established to make dyestuffs out of coal tars from day one, BASF held a strategic position in coal, which is reflected by the fact that the company in 1907 acquired a piece of coal mine assets in Marl, Germany (source of BASF website). Built on the success in making ammonia by fixing air-N BASF ventured on further to develop other synthetic technologies and necessary catalysts such as to turn syngas to methanol in 1923, another technology operated under similar pressure to the ammonia synthesis.

Industrial Synthesis for Liquid Fuels

Driven by the demand for liquid fuels, BASF then in next three decades ventured into manufacturing liquid fuels by turning coals to liquid fuels by hydrogenation, a technology invented in 1914 by Frederich Bergius (1884–1949) to break up the =C= bonding in coal at pressures of up to 700 bars. Bergius was not strange to the high pressure process system. He assisted Haber and Bosch in developing the ammonia process at the University of Karlsruhe in 1909. In 1927, BASF (part of IG Farben) commissioned the first hydrogenation plant to produce 100,000 tons of gasoline annually at Leuna in central Germany. The plant was later expanded to produce 650,000 tons of gasoline by processing brown coal and brown coal tar under a pressure of 250 bars (Storch, Hydrogenation of Coal and Tar, 1945). By 1939, there were total seven hydrogenation plants in operation with a total annual capacity of more than 2.2 million tons of liquid products. By 1943, five more plants were put into services, adding another two million tons of capacity into service. At peak time before the Allied strategic bombing in 1944, the twelve hydrogenation plants (Fig. 10.1) had reached an annual production rate of 3.25 million tons of liquid products including gasoline, diesel, lubricant oil, fuel oil and wax etc. Each plant had its own set of operating parameters with feedstocks ranging from bituminous coals, lignite coals, brown coals and their tars or tars made from other sources such as the carbonization process. The operating pressures of those plants vary from 250 to 700 bars.

The efforts to turn coal back into liquid fuels did not stop at the hydrogenation. Soon another catalytic process turning a water gas directly into liquid fuels followed the suit, the Fischer–Tropsch process or the indirect liquefaction. The hydrogenation process is also called the direct liquefaction.

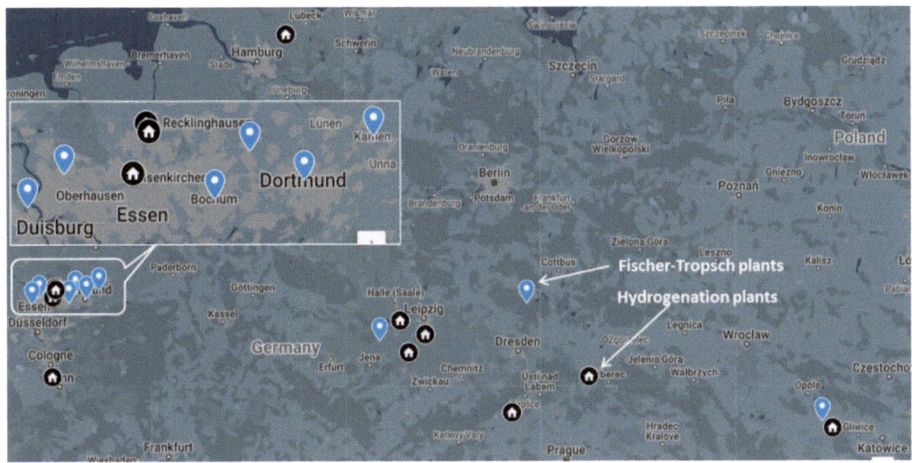

Fig. 10.1 Hydrogenation and Fischer–Tropsch plants developed in Germany

In 1933, the Ruhrchemie AG built the first demonstration facility with an annual capacity of 1,000 tons of motor fuel and lubricating oil in Oberhausen-Holten, northwest of Essen, Germany. By 1939, nine Fischer–Tropsch plants under the license arrangement with Ruhrchemie AG had been placed into service in Germany (Fig. 10.1). The total designed capacity from the nine plants reached somewhere near annual 600,000 tons of synthetic oils including gasoline, diesel and lubricating oils etc. At the peak time, the production from all nine plants had reached close to the design capacity. Under the license arrangement, Ruhrchemie AG would provide makeup catalysts to its licensees while retrieving the used catalyst for regeneration.

Driven by the thirst for liquid fuels and chemicals and enabled by technologies such as hydrogenation, Fischer–Tropsch, Harbor-Bosch for ammonia, and methanol synthesis etc. The demand for water gas had increased expeditiously, and therefore, had led to another round of extensive innovation of gasification technology.

References

Corp, T. S. (1980). *Coal gasification systems engineering analysis final report*. The SDM Corp.
From alchemy to chemistry. (2000). Retrieved August 2022, from University of Illinois at Urbana-Champaign: http://rbx-exhibit2000.scs.illinois.edu/cannizzaro.htm
Larter, J. E. (1907). *Producer gas*. Longmans, Green, and Co.
Morgan, J. (1945). Water gas. In H. H. Lowry (Ed.), *Utilization of coal chemistry* (p. 1673). Wiley.

Page, G. S. (1879). *Residual results* (pp. 399–417). American Gas Light Association.

Rambush, N. E. (1923). *Modern gas producers*. Benn Brothers Ltd.

Roehrig, W. C. (1870). *A practical treatise on metallurgy*. Longmans, Green and Co.

Storch, H. H. (1945). *Hydrogenation of coal and tar*. Wiley.

Technology, K.-O. E. (2019). *History of amonia plants and technology*.

The Making of Hydrogen and Ammonia

11

After WWI, the demand for liquid fuels such as aviation gasoline and motor fuel seems to go stronger than in the time prior to the war. In the meanwhile, the diesel engine that Diesel invented in the 1890s had finally become mature and started to be deployed by industries for different applications as well as a prime driver, which came right in to compete with the producer gas fueled gas engines. This in essence stalled further growth of the gas engine market. What is worse is the fact that the conventional coking facility for iron and steel industry also started to implement its operation by recovering the coke oven gas (COG) as a fuel for internal use and for export, which was pretty much wasted previously. The COG is a far better fuel gas than producer gas in terms of heating value. Hence in some ways, the development of industrial synthetic technologies provided a great opportunity for coal gas to stay relevant and to grow to serve a completely new industry that led to the establishment of the modern chemical and petrochemical industries. What is worth noting though is the fact that each of the emerging synthetic technologies from the synthetic ammonia, the methanol synthesis, the hydrogenation of coal to the Fischer–Tropsch process has its own set of specific and rigid requirements on what comes in with the syngas, hydrogen and carbon monoxide. It has to be of high purity and free from other impurities including tar, particulates, moisture, and the poisonous sulfur compounds and carbon oxides to the catalysts for some cases. For gasification to play a role in any of these syntheses at the time, therefore, it had to work on its own limits to make the necessary shift to a chemical play from the traditional fuel gas play. Otherwise, any gasification processes including water gas generator and gas producer would be impossible to deliver the syngas that would meet the specific requirements of each individual synthetic technology. After all, the emerging synthetic industry is a totally different playfield than a furnace firing, internal combustion or illumination. Different from the applications as fuel gas thus far, only two ingredients from coal gas are needed for the catalytic synthetic processes; they are hydrogen and carbon monoxide. Nitrogen could be the third one if producer gas is selected to feed an ammonia synthesis. All other components in coal gas such as the olefinians critical to illumination, the hydrocarbons important to firing up along with carbon dioxide and sulfur compounds have to be removed to their minimum. Failure to meet such requirements would render catalysts to lose its life. This time, the focus has turned to ways to recondition the syngas and subsequent purifications to remove other unwanted components so as to make the syngas applicable to the catalysts deployed with different technologies. For example, ammonia synthesis would require a volumetric ratio of hydrogen (H_2) to nitrogen (N_2) of 3 while poisonous components such as sulfur containing compounds, carbon monoxide and carbon dioxide have to be as little as possible, to the ppm level. The hydrogenation process seems a more tolerant process and hydrogen purity between 92 and 96% would help minimize the energy consumption by the extensive subsequent compression necessary for it. For the Fischer–Tropsch synthesis, however, sulfur compounds are the typical concerns that have to be minimized. The process also requires the syngas to have hydrogen to carbon monoxide ratio of 2 by volume. It, therefore, becomes obvious that additional technologies and processes are necessary to recondition the syngas and purify it to meet each of the individual synthetic processes.

Viable Industrial Hydrogen

By the early twentieth century there were a few hydrogen sources already commercially available for the production of hydrogen, which is in addition to the approach by the gasification of coal. Examples are the electrolysis to split water, the Lane process to break up water in steam, the COG from metallurgy coke making, and the by-product hydrogen from the chlorine-alkali electrolytic process and so on (Fig. 11.1). Among all these sources including gasification, however, only the hydrogen produced from an electrolyser or as a by-product from a chlorine-alkali electrolytic process is pure enough, for example, for ammonia synthesis. The gases originated from other sources would have to go through additional process steps such as reconditioning, separation, and purification etc. Let's take a look at the pros and cons of syngas making options in terms of syngas supply.

By 1900, the electrolysis process to split water had already been more than a 100 year old art; therefrom hydrogen could be generated at the cathode and oxygen from the anode by selecting the right electrolyte solution. However, its industrial application had not started until the Russian engineer Dmitry Lachinov (1842–1902) developed a viable design for an electrolyser in 1888, a time when a steady supply of electricity became available. Although there were a few hundred devices in industrial use by the time of his death the scale was typically small considering the high price of electricity. What is attractive with the electrolysis though is that the hydrogen generated therefrom is of high purity, which could be utilized directly as a feed to the synthetic ammonia process. Except

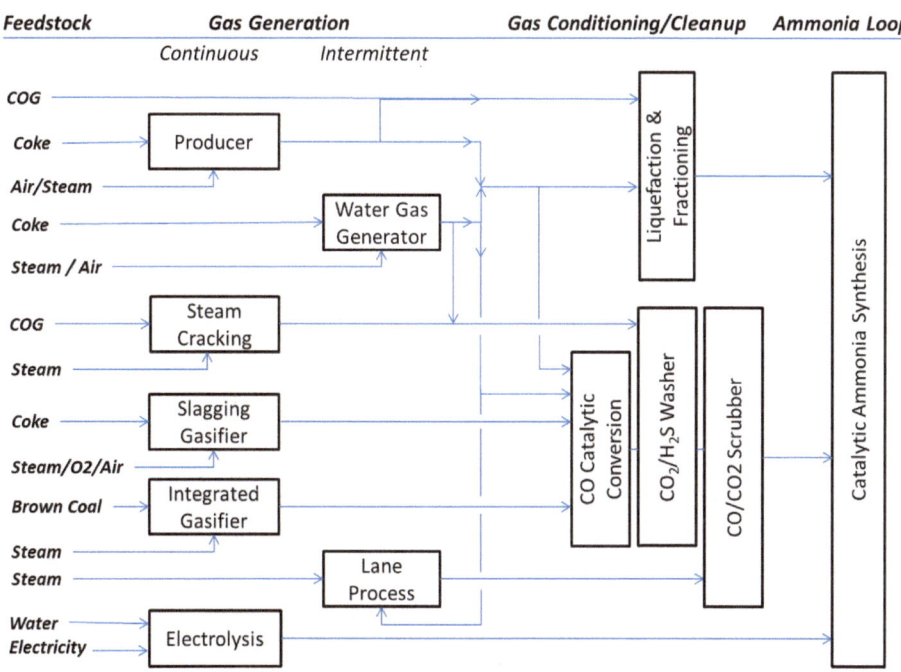

Fig. 11.1 Commercial processes for hydrogen production

in locations or places where cheap hydro-electricity is available such as the Niagara Falls and Sweden, such hydrogen would hardly be affordable for the production of ammonia even at a small scale. With regard to the by-product hydrogen generated from a chlorine-alkali electrolytic process, it is worthwhile to mention in a context that the chlorine-alkali electrolytic process is a technology developed by the American chemist and engineer, Hamilton Castner, and the Austrian chemist, Carl Kellner. The two inventors then formed the Castner-Kellner Company in 1895 to manufacture caustic soda and chlorine, which came to compete directly with Solvay products. The company was, however, taken over by Brunner, Mond and Co. in 1920. So the Solvay process itself could not bypass the severe competition and ended with a short life soon. The chlorine-alkali electrolytic process has since remained dominant in the chlorine-alkali industry till this day. In terms of hydrogen production, the high price of electricity would seem a lesser issue than the fact that hydrogen as a by-product was subject to the production of the major products. Except for niche cases, however, hydrogen production via electrolysis would hardly be practical in general for any large scale deployment like the emerging synthetic processes.

The Lane process was a thermal process invented by the British engineer Howard Lane in 1903. Similar to the blue water gas generator, it is an intermittent process by passing steam through a hot reduced iron bed to generate hydrogen; the iron once becoming oxidized by steam would need to be reduced again through a redox cycle of iron oxides.

To regenerate the oxidized iron a water gas is typically used to reduce the oxidized iron back to its reduced state. To start the cycle operation, the water gas is also used to heat up the iron bed to the required temperature before switching to the cycle operation. With multiple generators working alternately or in rotation, a continuous supply of hydrogen could be made possible. If a careful operation procedure is practiced the generated hydrogen could be of very high purity. By 1909, the Lane process unit could generate up to 10,000 cubic feet of hydrogen hourly, which was often used to inflate airships, balloons, and to make special organic chemicals such as hydrogenated vegetable oils (Lane, 1909). For a synthetic ammonia application, however, additional purification to remove a small amount of carbon oxides would be necessary, which more follows.

At the time when BASF was developing the industrial process for ammonia synthesis the only practical source of a reliable hydrogen supply in a large quantity was clearly the syngas produced via a water gas generator considering the facts that there exists an unlimited supply of coals and that the water gas making had become mature enough to provide an economical syngas. A coke oven gas could always be used as a supplement for additional hydrogen, whenever available, from a nearby metallurgy coke oven operation or a low temperature carbonization plant. The COG typically contains 50% H_2, 15–20% N_2, 20–30% CH_4, unsaturated hydrocarbons and a small amount of carbon monoxide. When getting cracked with steam, the COG would be an excellent source of hydrogen, which is why the COG steam cracking had been often used for many of the future Fischer–Tropsch operations. For other synthetic processes such as the methanol synthesis, the hydrogenation of coal and the Fischer–Tropsch synthesis, however, the producer gas would be out of the picture due to its high nitrogen content. Then the question becomes how to make water gas and coke oven gas fit the respective synthetic processes in terms of purity and composition. For example, the desired gas for the hydrogenation of coal would be high purity hydrogen, for ammonia synthesis hydrogen and nitrogen with a volume ratio of 3:1, and for synthetic methanol or a Fischer–Tropsch liquids hydrogen and carbon monoxide with a volume ratio of 2:1. To get there, carbon monoxide in water gas would have to be removed or reduced depending on the synthetic process. This is similar to the challenge that Mond and Langer faced back in mid-1880 in their efforts to bump up the content of hydrogen in the Mond gas by catalytically reducing others such as carbon monoxide, carbon dioxide, and many other hydrocarbons. What is different around 1910 though is that there was a technology that had become available for the purpose to separate the different gases in a water gas based on the difference in boiling points of each individual gas (Figs. 11.1 and 11.2), that is, the liquefaction and fractioning or the cryogenic process.

Liquefaction and Fractioning

While being recognized for his work on energy equivalence and transformation, Joule continued his enthusiasm for scientific experiments. Around the mid-1850s, his collaboration with Lord Kelvin made another discovery. In their experiments with gases, they found out that when a gas at a pressure expands freely into a vacuum space the temperature of the gas drops. Further experiments demonstrated that a drop in one atmospheric pressure would cause a decrease in the air temperature by roughly a quarter °C. The finding was later explained by the Dutch physicist, Johannes D. van der Waals (1837–1923), with his molecular kinetic theory that the temperature of a gas is proportional to a square of the average velocity of the gas molecules. When the velocity of gas molecules slows down so is the temperature of the molecules, which has become the so called Joule–Thomson effect. The Joule–Thomson effect then led to a few important innovations for gas handling, refrigeration, and oxygen separation etc. during the early twentieth century.

The Joule–Thomson effect had been exploited extensively both for scientific purpose and for industrial deployment, which both are important for the advancement of coal gasification technology. In the former field, Louis-Paul Cailletet (1832–1913), a French physicist and ironmaster, is a noted individual for his work on the liquefaction and isolation of gases. Cailletet was the first to liquefy oxygen in 1877, and many other gases such as nitrogen dioxide, carbon monoxide, and acetylene. Cailletet was also the one who was able to finally achieve the liquefaction of nitrogen and hydrogen in 1898, which was carried out independently as well by the Swiss physician Raoul-Pierre Pictet (1846–1929). By looking at the boiling point of the gases in Fig. 11.2, liquefying hydrogen would need a temperature of as low as −252.9 °C, which tells why it had taken such a long time to achieve it. The separation of individual gases had allowed chemists and physicists to study those individual gases for their behaviors, characteristics and thermodynamic information. Such information along with the establishment of the molecular theory has made a quantitative performance of those many **Reactions (1–7)** in Table 10.1 of Chap. 10 taking place in a producer or a water gas generator possible. It also makes the performance estimate of gasification possible under certain assumptions as did by Dowson, Rambush and Gumz et al. This laid a foundation for future attempts to mathematically simulate the gasification process via certain kinetic modeling.

On the industrial side, Thaddeus Lowe after leaving the US Balloon Corp. in 1863 invented an ice making machine by using carbon dioxide as a coolant in a closed loop to extract the cold heat by freezing water to make ice for refrigeration. In the Letters patent he filed in 1867 (Lowe, 1867), Lowe described his invention with *"I employ in my process carbonic acid gas, first condensed by mechanical pressure to the form of a liquid, and then, the pressure being withdrawn, allowed to resume its normal state by expansion...."* Although his attempts to apply the new refrigeration process to shipping fresh beef from the Midwestern US to New York City did not go as planned Lowe certainly becomes one of the early few who tried to make use of the newly discovered scientific principle,

Fig. 11.2 Boiling points of gases, °C

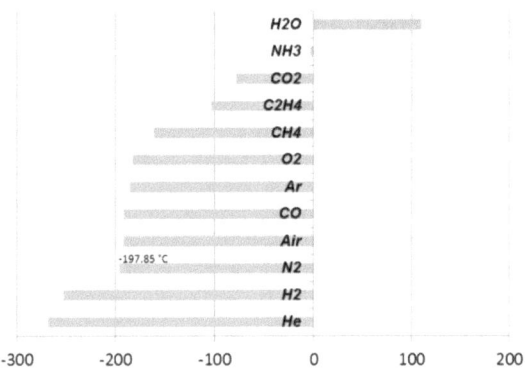

the Joule–Thomson effect. Soon after, the refrigeration with ice was cut short by another technology developed in Germany. It is refrigeration by using ammonia as coolant, which is more efficient than using ice. The technology was invented by Carl von Linde (1842–1934), a German chemist and engineer, who received his education at the Swiss Federal Institute of Technology in Zürich, Switzerland, during which von Linde was exposed to the teaching by several influential advisors including Rudolf Clausius. After a few stops at different shops and factories, von Linde landed on a position as a lecturer at Technische Hochschule of Munich in 1868 and became a Professor four years later with a focus on heat theory. It is during his professorship von Linde made his invention by making ammonia a better coolant for refrigeration between 1873 and 1877. Then, von Linde set up his company, the Linde AG, in 1879 to make and sell this new type of refrigeration machines and gas products. The success was immediate (Praxair website) (German Culture). The next technology that von Linde invented is much more significant and has a long lasting impact on many fields and industries including gasification till this day.

In 1895, von Linde compressed air, cooled it down and let it expand freely through an expansion valve. When the temperature of the expanded air drops further the air eventually becomes liquid. Then, by warming up the liquid slowly von Linde succeeded in separating the air into nitrogen and oxygen. At the Paris World Fair in the same year, von Linde showcased the first air separation system, processing 3 L of liquid air by compressing air to 200 atmospheres and then letting it down to 30 atmospheres (Linde website). In 1902, von Linde built the first single column air separation unit that was capable of producing pure oxygen. Concerning nitrogen, however, the system could only produce a stream of nitrogen containing about 7% oxygen, which would not be suitable for ammonia synthesis. Around the same time, George Claude (1870–1960), a French engineer, improved the refrigeration cycle by adding an expansion engine and a mechanical brake to reduce the enthalpy of the air vapor, and therefore its temperature, at nearly constant entropy. Such an improvement by applying the principle of an isentropic expansion greatly reduced the cooling time and operating pressures of the liquefied air while increasing its throughput of it. As a result, Air Liquide was born in Paris in 1902 (Cryogenic Society website). Such an air separation technology or the air separation unit (ASU) has become one of the most important industrial processes; the oxygen thus available soon began to change the making of coal gas or syngas (Fig. 11.1).

Vying to the strong demand for oxygen in metal cutting and the ongoing industrial development led by BASF for ammonia synthesis, von Linde in 1910 improved his single column design to a double column by adding a low pressure column on top of the pressure one. Such an improved system became capable of producing both oxygen and nitrogen of a high purity, which the latter could be used directly in ammonia synthesis by combining with three volumes of hydrogen for each volume of nitrogen. The early deployment of ASU for coal gasification started around the mid-1920s at the Leuna industrial complex. When the Linde-Frankl design became available more ASU were deployed at many of the synthetic plants to provide oxygen to the gasification operations. At the Leuna plant, the first two Linde-Frankl units were put into service in 1928 with a unit capacity of 2,000 m^3/h, and then the number of units increased to six by 1940 producing 50,000 m^3/h of nitrogen with a purity of 99.5%, which was used as a protective gas and for the ammonia synthesis. The balance of oxygen was consumed by gasification operations. There was also krypton gas recovered as a by-product. To meet the increasing onsite demand for oxygen another unit was put into service in 1941 to provide additional 3,560 m^3/h of oxygen of 98% purity (Plant, 1940). By 1943, there were about 43 Linde-Frankl units with a unit capacity designed from 1,000 to 4,000 m^3/h up running to provide 126,000 m^3/h of oxygen in eight synthetic operating facilities in German territory at the time (Allied Investigation Mission, 1945; Power, 1947). In international market, Linde set up his US licensing arm, the Linde Air Products Company, to conduct an international business in 1907. After a few changes the company in 1992 became an independent US company, the Praxair. Technologically, the double column design of ASU has remained essentially unchanged for more than a century till today except that its scale has increased to a much larger capacity on a unit base, producing up to 120,000 m^3 of oxygen per hour for some of the coal gasification plants developed in places such as China in the past ten years.

Technically speaking, the same principle as with the air separation by applying the liquefaction and fractioning can be used to separate hydrogen from syngas or COG on an industrial scale as well. With the fact that hydrogen and nitrogen have the lowest boiling points among other gases (Fig. 11.1) a high purity hydrogen or hydrogen and nitrogen could be obtained to feed directly into the ammonia loop via a cryogenic approach. By taking the advantage of the high nitrogen concentration in producer gas, blending it with either COG or water gas in the right proportion and then liquefying and fractioning the blended stream would allow the production of the right mix of hydrogen and nitrogen of 3–1 for the ammonia loop, avoiding of the use of nitrogen from an ASU. Under such a principle, the Claude process was one of the early processes used, when necessary, to separate hydrogen, nitrogen or carbon monoxide from water gas, producer gas or COG.

In such an operation, the water gas or the COG, free from tars, ammonia, oils, benzene and sulfur, is scrubbed at high pressure with water and caustic, then treated by cooling it with ammonia to -45 °C to remove water, after which it goes to the liquefaction and fractionation process, where the process is similar to the separation of N_2 and O_2 in ASU. In general, COG usually works well for ammonia synthesis due to its N_2 content while a water gas is more suitable for the production of hydrogen and carbon monoxide. An example is that an ammonia plant located at the Castrop-Rauxel, Germany deployed five Claude refrigeration units and one of another type to separate and purify hydrogen from COG as its feed gas to manufacture about 180 tons of ammonia daily. The plant was a neighbor to the second Fischer–Tropsch plant built at Castrop-Rauxel in 1936. Both the plants had been in operation until being bombed by the Allied forces in June 1944 (Powell, 1945).

More often than not, the decision between choosing the Claude process and its alternative for gas separation depends on which process would provide the desired economic benefits. A problem with the Claude process hinges on the fact that obtaining a temperature of -197 °C or even lower requires a significant amount of compression energy, which often requires a sophisticated refrigeration system. Except for some special occasions like the example mentioned above or a small scale use in laboratories for scientific purposes, the use of liquefaction and fractioning had been limited. The need for viable alternatives was still wide open for the separation and purification of syngas. Taking the ammonia production at Oppau as an example it adopted a blend of water gas and producer gas as the source of hydrogen and nitrogen, but did not use the cryogenic approach to purify the blended gas that was fed into the ammonia loop by the time of 1921. Instead, it used a chemical approach, somehow similar to what Mond and Langer did back in the mid-1880s.

Syngas Conditioning and Purification

From the patents filed by BASF during the early 1910s, Bosch might have most likely realized that coal gasification would be the only reliable way to a secure the supply of hydrogen while with an ample quantity. Obviously, the needs for a viable method or process to separate and purify hydrogen out of a water gas became real and urgent while working on the design and construction of the first ammonia plant. As Professor Childitch of the Industrial Chemistry of the University of Liverpool stated in 1929 (Childitch, 1929) that,

> This company (BASF) set itself, with characteristic thoroughness, to work out technical means of catalysing the water–gas reaction, and, in a long series of patents, published a whole variety of catalysts, temperatures and pressures by means of which it was said the process could be successfully carried out. In their first patents they emphasised the use of catalysts containing nickel, cobalt, and mixtures of similar metals, but it was evidently soon found that

catalysts of this type had too active tendencies towards hydrogenation, with the result that an undue amount of the carbon monoxide present was transformed into methane, which would serve merely as a useless diluent of a gas mixture intended for use in synthesizing ammonia.

Bosch must have had a comprehensive plan to tackle the issue of a reliable hydrogen supply for ammonia synthesis, which would pave a solid foundation for other initiatives such as methanol synthesis and coal hydrogenation to liquid fuels as well. The water gas made from coke contains about 51% hydrogen, 40% of carbon monoxide, 4% carbon dioxide, 5% nitrogen, and a trace amount of organic sulfur and hydrogen sulfide. To best use the water gas, the logical way would be to convert carbon monoxide with steam to hydrogen and carbon dioxide and then to remove carbon dioxide and other impurities including carbon monoxide, carbon dioxide, and sulfur containing compounds so that the remaining hydrogen becomes purified to meet the requirement of ammonia loop. To convert carbon monoxide to hydrogen, what Bosch did first is to pick up the work left by Mond and Langer in the mid-1880s by using nickel and cobalt as catalysts to turn carbon monoxide to hydrogen by reacting with steam. Carbon dioxide would form as well.

In retrospect with regard to converting carbon monoxide over the catalyst back then in their efforts to prepare hydrogen for fuelcell use, Mond and Langer might have not realized the role and contribution of steam in the system. Based on their observation of increased contents of both hydrogen and carbon dioxide when introducing the Mond gas along with steam into the catalytic converter they believed that carbon monoxide gets cracked via the Boudouard reaction ($2CO -> C + CO_2$) over the catalyst to carbon dioxide and carbon, where the latter was the deposit they observed on the catalyst. The significant increases in hydrogen to 36–40% from 25% would, however, be hardly explained by the cracking reactions of CO and HCs under the conditions. In reality, such a carbon deposit could be from the decomposition of other hydrocarbons in the Mond gas. The reduction of carbon monoxide and the increases of hydrogen and carbon dioxide could well be due to the reaction of carbon monoxide with steam (**Reaction 6** in Table 10.1) over the cobalt or nickel as catalyst. Remember that the Mond gas directly from the Mond gas producer carries a large amount of steam. After all, this part of discussion might not be the focus of Mond and Langer at the time. By the beginning of the next century, however, the reaction of carbon monoxide and steam, its equilibrium nature, and temperature dependence had obviously been investigated by Dowson and presented in his 1907 book (Larter, 1907). Actually, the pioneering work started by Mond and Langer had been followed by many other inventors in the subsequent years. All these inventions, unfortunately, had not been able to generate meaningful results for industrial use until Bosch and his colleagues laid their hands on the subject. Based on multiple patents that BASF filed during 1912 and 1914, Bosch had finally landed on a catalyst of an iron oxide promoted with alumina or other similar metal oxides that successfully converted carbon monoxide to hydrogen and carbon dioxide under the conditions similar to those used by Mond and Langer back then, a temperature between 450 and 500 °C, about 50° higher than what Mond and Langer used. Under normal operating conditions, the catalytic converter, often in two stages when necessary, could reduce carbon monoxide to below 2%.

With the success in converting carbon monoxide into hydrogen, an equal volume of carbon dioxide forms as well. To remove the carbon dioxide, BASF developed an absorption process by using water washing at pressure and it worked well. At this point, the treated gas contains almost hydrogen and nitrogen plus a small amount of residual carbon monoxide and carbon dioxide. The residual carbon monoxide and carbon dioxide are still poisonous to the catalyst and therefore need to be removed to minimum. To do so, BASF then invented a scrubbing process by using an ammoniacal solution of cuprous salts such as copper formate at high pressure, which would effectively remove remaining carbon oxides to the level that satisfies the catalyst used in the ammonia synthesis. The spent ammoniacal solution will be recovered by letting down the pressure to release the absorbed carbon oxides for recycle. With regard to any residual sulfur containing compounds such as hydrogen sulfide and organic sulfur depending on its source, they are poisonous to the catalyst as well and need removal. Coincidently, the organic sulfur would be converted to hydrogen sulfide along with carbon monoxide in the catalytic converter. Then hydrogen sulfide would be washed away along with carbon dioxide by water washing. What a synergy of the process.

All these developments, together with all the experiences acquired with coal gasification, either gas producer or water gas generator, had been built into the first ammonia plant at Oppau.

Hydrogen Production at Oppau

Being commissioned during the fall of 1913, the Oppau ammonia facility was built on an abandoned Leblanc soda works of BASF. By 1921, the ammonia capacity at Oppau is estimated at around 65,000 tons annually, a significant increase from the annual 9,000 tons at the beginning. The process for hydrogen production is reflected in Fig. 11.3 where both the producer gas and the water gas were respectively processed to free tars, liquors, and dust by the means of a hydraulic main and a water washing, similarly practiced with coal gas making, from the respective gases before being metered and blended into syngas, a mixture of hydrogen and nitrogen with a volume ratio of 3/1. The syngas then goes through additional cleanups, both chemically and physically, in order to meet the requirements for an ammonia synthesis (Childitch, 1929; Parker, 1923; The Making of Oppau, 1921). Here is a brief description of the major process flow that was built into the Oppau plant by 1921.

There are eighteen water gas generators grouped into nine bundles, each bundle shares a common control board, a scrubber and a washing tower. The generators are of the conventional Humphrey and Glasgow type, a cylindrical sheet iron vessel lined with

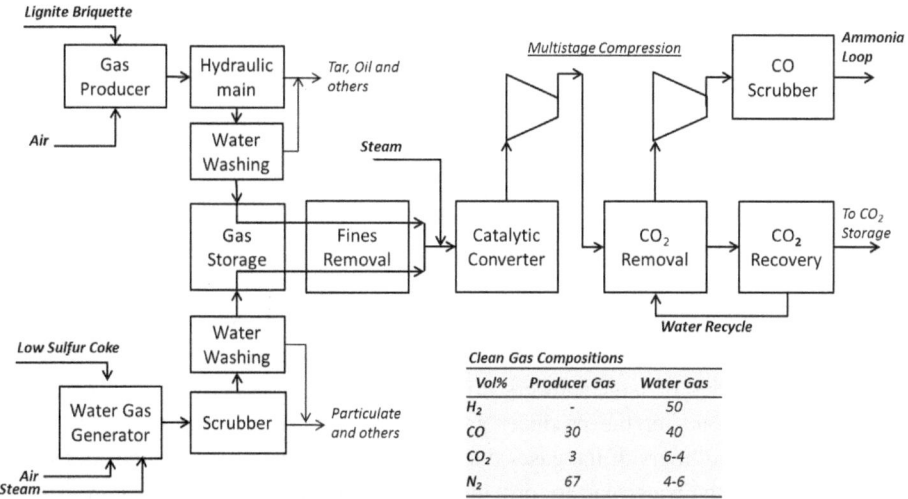

Fig. 11.3 Gas making at Oppau for ammonia synthesis (*Note* The producer gas should contain a certain level of hydrogen with a lignite coal as feedstock.)

refractory bricks and equipped with mechanical grates. The generators fed with a local low sulfur coke are operated in a Blast Run cycle of 4 min as below:

1 min: air Blast from the grate

½ min: purging with superheated steam from the grate

2.25 min: Run with SH steam from the top

2.25 min: Run with SH steam from the top

Then repeat the air blast to start a new cycle.

The water gas generated from the Run operations enters the washing towers for the cleanup of tars, oils and particulates, after which it is collected in two gasholders for a short time of storage.

Lignite coal from the Cologne area was made into briquettes that were used for the production of producer gas as a source of nitrogen for ammonia synthesis. Producers of at least twenty-one units are housed in the same building as the water gas generators, and grouped into three bundles, seven in a bundle. The generators are of cylindrical sheet iron design lined with refractory bricks as well. Due to the use of lignite coal, the cleanup of the producer gas was more complicated than the water gas produced from a low sulfur coke, which is why the producer gas has to go through an extensive cleanup processes including hydraulic main, a series of washing towers, purifying turbine type rotating washers, and additional cyclones to remove tars, oils, dust and ammoniacal liquor. The departing gas is then stored in a gasholder as a reservoir for a short period of time.

The control boards were provided with manometers for pressure monitoring, which were linked to the regulating valves for the operation of the water gas generators and the gas producers.

During normal operations, the producer complex consumes a total of 440 tons of lignite briquettes and gives off 1,235,000 cubic meters of producer gas daily and additional 10 tons of tars as a by-product. Only about 15% of the producer gas goes to blend with the water gas for the ammonia loop, and the rest is consumed as fuel gas around the plant. The water gas generators consume a total of 265 tons of coke, and generate 375,000 m^3 of water gas every twenty-four hours. The required air for the water gas generators is supplied via a series of large blowers driven by electric motors. The three gasholders, 15,000 m^3 each, one for the producer gas and two for the water gas, would store up gases for about two hours of the gases during normal operation. The compositions of the producer gas and the water gas are provided in Fig. 11.3 and their qualities are regularly monitored by measuring the concentration of carbon dioxide and heating values of the gases, respectively.

To minimize any potential impact on downstream catalytic conversion, the producer gas and the water gas are drawn from the gasholders to pass through six turbine type washers, four for the water gas and two for the producer gas, in order to remove any residual particulates in the gases. Then, the clean gases are routed to three rotary meters, two for the water gas and one for the producer gas, and metered in two to one ratio in volume to merge into one stream for the chemical treatment in downstream process units such as catalytic CO conversion, CO_2 removal, and additional cleanup of CO and CO_2. At this point, the mixture of hydrogen and nitrogen has finally become ready as the raw material gas for the ammonia loop to make ammonia, typically containing no more than 500 ppm carbon monoxide and about 1 ppm sulfur compounds.

In an actual operation, the ratio of hydrogen to nitrogen was typically maintained a little higher than 3–1 during the metering and mixing of water gas and producer gas. The accurate ratio could be maintained, when necessary, conveniently by injecting the balance of nitrogen from an onsite ASU prior to the ammonia synthesis loop. If it were otherwise there would be no easy source of high purity hydrogen available at the time.

With the success achieved at Oppau, BASF soon opened ground for the second ammonia plant at Leuna, central Germany in May 1916 and completed it about 12 months later. In 1918, the two plants churned out about 219,000 tons of ammonia, equivalent to 850,000 tons of ammonia sulfate, a quantity sufficient enough to replace Germany's Chile nitrate import, which was about half of the production of the Chile nitrate for the same year. By 1920, the Haber–Bosch process had made up almost 20% of the total nitrogen-derivative fertilizers worldwide, a jump from almost none prior to the ammonia plant at Oppau. During the same period, the Chile nitrate import reduced to almost half and the by-product ammonia sulfate recovered from both the coal gas and the coke oven gas dropped to 27% from 38%.

During the war time, the British government set up a committee, *the Nitrogen Products Committee*, recommended by the Faraday Society to investigate the Haber–Bosch process for ammonia making. The Committee made an estimate based on the available information on the production cost at Oppau. It suggests that it would cost 15.57 Pounds to produce a metric ton of ammonia before applying interest on capital to it (Table 11.1). For comparison, nitric acid was generally produced from the Chile nitrate. The cost of nitric acid produced from the Haber–Bosch ammonia was about 50% of that produced from the Chile nitrate before the war. The synthetic ammonia was also much more competitive to the ammonia sulfate recovered from the operation of the coal gasworks on an ammonia basis. It is interesting to note that hydrogen production clearly makes up almost 62% of the total cost of ammonia production. Considering the fact that the source gases, either water gas or producer gas, only costed between 4 and 12d per 1,000 cubic feet before the war, the resulting syngas for ammonia loop would bear a price tag of 2s 6d per 1,000 cubic feet, which is obviously contributed from the additional cleanup processes, both chemical and physical, and the energy consuming compressions. The syngas cleanup and purification is for sure a costive process.

The significance of the Haber–Bosch process comes in many ways. First of all, the ultimately man-made fertilizers on a large industrial scale enabled a revolution in agriculture that has provided the world with ample food to feed the fast growing population ever since. Ammonia production increased from annual production of a few thousand tons from the first plant at Oppau in 1914 to an astronomical 150 million tons in 2021. The technology and processes so developed and the hands on industrial experience acquired

Table 11.1 Cost of ammonia production by Haber–Bosch process in 1922 (The Nitrogen Industry 1922)

Cost items	Rate	Unit	Unit price	Cost (£/ton)
Hydrogen for ammonia synthesis	77,000	1,000 cubic feet	2s 6d	9.63
Nitrogen for ammonia synthesis	26,000	1,000 cubic feet	6d	0.65
Electricity for N_2 compression and catalyst heating	1,500	kW h	0.25d	1.56
O&M costs (wages, superintendence, general expenses and fixed charges)				1.89
Depreciation on catalyst and plant				1.84
Total				15.57

to engineer, operate, and maintain the synthetic process of the high pressure and high temperature rolled directly into the commercialization of the coal hydrogenation process, the methanol synthesis, and the Fischer–Tropsch process, only with a bit twist in the downstream process, which subsequently created a significant demand for water gas. The Nobel Prizes awarded to Haber in 1919 and later in 1931 to Bosch and Bergius is a simple recognition of such significance.

From the coal gas making perspective, the development of the catalytic CO converter is an important step that made water gas an affordable source for the industrial production of chemical grade hydrogen and carbon monoxide, which are important commodities for modern chemical and petrochemical industries. The catalytic CO conversion itself has become one of the most important industrial innovations as well, known as the Water Gas Shift Reaction (WGSR) process.

Coal gasification has now entered into a completely new territory, the age of chemical synthesis.

References

Allied Investigation Mission, W. (1945). *Report on investigations by fuels and lubricants teams at the I. G. Farbenindustrie A. G. Works at Leuna.*

Childitch, T. H. (1929). *Catalytic processes in applied chemistry.* D. van Nostrand Co.

Lane, H. (1909, August 28). The Lane hydrogen producer. *Flight*, 524.

Larter, J. E. (1907). *Producer gas.* Longmana, Green, and Co.

Lowe, T. (1867). *Patent No. 63,404.*

Parker, J. R. (1923). *The nitrogen industry.* Constable & Company Ltd.

Plant, L. (1940). *Annual report 1940 of ammonia plant Merseburg.*

Powell, C. H. (1945). *Plant of Klocknerwerke, AG—Castrop-Rauxel, Germany.* Office of Publication Board, Department of Commerce.

Power, M. O. (1947). *Report on the petroleum and synthetic oil industry of Germany.* The Ministry of Fuel and Power.

The Making of Oppau. (1921). *A weekly technical newspaper, being the incorporation of electro-chemical and metallurgical industry and iron and steel magazine* (p. 305).

Reinventing the Coal Gas Making

12

Post WWI, the shift in the coal gas market continued. On the one hand, the electricity generation market seems to have increasingly firmed up its establishment, kicking the coal gas illuminating business out of the cities and eventually towns, which appears a sure thing to happen. On the other hand, the producer gas driven gas engines had apparently reached the upper ceiling because the fields of gas engines for either electricity generation or as a prime driver had probably reached its peak due to several developments in the market place. First, the operation of the generation stations had become more complicated by frequently running at different loads to match the demand change of electricity on a daily basis. Gas engines tend to be less efficient during turndown operations than steam engines. Secondly, gas engines had become its own victim of going large in size in its competition with the steam engines. When gas engine size goes beyond a certain level, say above 1,000 HP, the piston or reciprocating gas engines would normally need to adopt a multi-cylinder design due to the size limit of the cylinder sizes. These multi-cylinders are arranged either vertically or horizontally. The former design was popular in Britain for gas engines up to 1,500 HP, and the latter in Germany often operated in tandem for up to 4,000 or 5,000 HP. Such large gas engines, however, tend to cost more to build, operate, and maintain; they also occupy a more floor space, making them less competitive against steam engines. And thirdly, a large gas engine would require more consistent and steady supply of a producer gas in order to maintain an optimum engine performance, which would place more constraints on the operation of a gas producer. With more experiences acquired it had become clear that the benefits of a large gas engine over a certain level would be adversely impacted. This essentially translates to a situation that the growth prospects in the gas engine field for the gas producers had basically become capped. In a certain way, the emerging synthetic industry provided a much needed space for both producer gas and water gas as the important raw materials, syngas, in any of the synthetic processes including the ammonia synthesis, the methanol synthesis, the hydrogenation of coals and the Fischer–Tropsch synthesis. To meet such a huge demand for syngas, however, gasification would need naturally to go larger in terms of unit capacity in order to reduce the costs to build, operate, and maintain. As an example, each of the water gas generators at the Oppau ammonia plant processes about 14.7 tons of coke daily assuming all the eighteen generators are in operation at the same time, no spare considered. Should the unit capacity of the water gas generator double the number of the generators could be reduced by half, and the cost to build, operate and maintain the plant could significantly be expected to go down; so does the cost of the syngas produced. The benefits with a large capacity gasifier are obvious.

The overall process of the Leuna ammonia plant followed primarily the footprint of the Oppau plant except for some minor changes made here and there. It was a much larger plant with a design capacity of 400 tons of ammonia daily. There were thirty-one gasifiers on site, among which 26 are water gas generators and 5 gas producers. These gasifiers are of a similar design, the Pintsch type (based on the old H&G water gas generator) with a large ID of 3 m. The plant capacity was soon more than doubled during war time. By 1918, the combined production of ammonia from the two plants reached 219,000 tons. Taking out 65,000 tons by the Oppau operation, the production rate at the Leuna plant should be somewhere about 154,000 tons of ammonia that year (Parker, 1923). According to the reports by the Allied Mission at the end of WWII, the central region of Germany lacks a good quality coals but is rich in lignite reserves, which is of high moisture and volatile matter. Therefore, the feedstock to make the water gas was typically oven cokes manufactured from bituminous coals in Ruhr district, which was railed about 280 miles to the Leuna plant (Holroyd, 1946).

The water gas generators installed early on at the Leuna ammonia plant were near the largest at the time. Their operations during the earlier years appear to have experienced a severe problem with clinkering of the coke used that causes a built up of ash at the bottom of gasifiers, preventing the generators from normal operation at the desired temperature. Obviously, the desired operating temperature seems high for the coke adopted and reducing the operating temperature would be a simple solution for sure to avoid the problem. This would, however, create another problem by impacting the production of the water gas and producer gas adversely, resulting in an ash containing more than 50% of carbon. The Leuna operation decided to lower the gasification temperature so to minimize the clinkering problem because the unreacted carbon in the ash would ease the clinkering formation so to make the ash discharging less problematic. It worked. Then utilizing such an ash containing such a high carbon ("refuse") became the new challenge. What the Leuna plant did was to build a new type of gasifier called the Wuerth gasifier, a slagging gasifier, which is operated at a higher temperature than the ash melting temperature so that the ash is discharged as a molten slag. What is unique about this slagging Wuerth gasifier is its continuous production of water gas, which is enabled by utilizing oxygen and steam as blast. Using oxygen instead of air as a blast opened up a new era for coal gasification to transform itself. The adoption of oxygen with a producer operation starts to make the border line between a water gas generator and a gas producer blurry, enabling a water gas generator to operate continuously in the making of water gas or syngas.

Slagging Gasification

The slagging Wuerth gasifier is essentially no difference in its operating principle than what Ebelmen demonstrated at the iron works in 1939–1940. Driven primarily by the gas engines of high power outputs, the slagging producers of different designs reemerged and found their occasional uses in Germany, the US, France, and Britain around 1900 to provide a hydrogen lean fuel gas for large gas engines in order to prevent a premature ignition or misfire during the compression cycle. Another driver for a slagging gasifier seems to be associated with achieving high throughput so to make the producer gas more competitive in order to better compete with steam engines in the market place. Fundamentally speaking, there are only limited ways to increase the throughput of a producer or a generator, and here they are

(a) Increase the size of a producer/generator.
(b) Select good quality and reactive coal and coke.
(c) Increase the temperature of gasification above the ash melting point.

In spite of the demand for large throughput gasifiers, water gas generators and producers, the harsh working environment for operators needs to be addressed. The operations of for either gasifier need a frequent tending to ash removal, which had been operated manually. It was a laborious job and required skill as well. Such a work would become hazardous when the size of a generator or a producer was larger than, say 2.8 or 3 m and above. From operation viewpoint, slagging gasifiers would address well the ash removal as the ash once becoming melted would flow out automatically as long as the environment is kept hot enough so as not to "freeze" the molten ash. Considering the largely available coke from coal gasworks, by-product coke ovens and many low temperature carbonization operations at the time, a slagging gasifier would seem a perfect solution to provide a hydrogen lean fuel gas while being capable of achieving a higher throughput because the cokes, coals, and any hydrocarbons would burn much faster at higher temperatures. Slagging gasifiers, therefore, were exploited in Germany during WWI and met with a certain level of success. A few different types of slagging gasifiers were developed by companies like the Wuerth, the Georgs-Marienhuette, the Rehmann, and the Julius Pintsch etc. and deployed in iron and steel works to drive gas engines. It is also applied to the chemical industry for the production of chemicals such as formic acid which requires carbon monoxide as an input material.

According to the report by the Allied Mission, the Leuna plant had already deployed three slagging Wuerth producers, brick-lined from top to bottom, to digest a refuse with air blast during the early days. The producer gas was sent to the ammonia loop along with water gas. It seems to be working but with a catch; the higher temperature caused a severe wear to bricks at the lower part inside the producers. To lessen the new problem, Leuna adopted six modified slagging Wuerth gasifiers, five supplied by IG Farben and one by Pintsch. The modification was to use a water jacket in the lower part instead of brickwork. Such a modification might have fixed the issue of brick wearing but created another one. The water jacket cooling draws too much heat so that it is difficult to maintain the necessary slagging temperature. This is because of the refuse has a low heating value and could not generate enough heat to raise the temperature high enough for a slagging operation while the resulting producer gas would be too low in heating value as fuel gas. This had probably forced Leuna to switch the air blast to an oxygen blast, which addressed the problem. The upgraded slagging Wuerth gasifiers, each capable of producing 15,000 m^3/h, would provide a total of 90,000 m^3/h of producer gas, which the producer gas typically contains little nitrogen and a combined carbon monoxide and hydrogen of 89.4% as shown in Table 12.1, which is well suited for chemical synthesis.

Prior to 1938, a total of nine ASU units of the Linde-Frankel type provided the required oxygen of 98% purity to the slagging operation. Two of the units are old, each with a unit capacity of 1,900 m^3/h and the seven new with an increased capacity of 2,875 m^3/h each. The Allied report recorded that the actual production of oxygen between 1938 and 1942 was 22,000 m^3/h, which is barely to fully load five of the slagging Wuerth gasifiers. Besides, ASU is an energy guzzler, each cubic meter of oxygen drawing about 0.55 kWh of electricity, which makes up 55% of the cost of oxygen assuming the electricity price running at 1.1–1.2 pfg/kwh at the end of the war.

Oxygen was expensive back then and is still not cheap in today's market. It is a special and necessary commodity that would transform the gasification technology in many decades later.

The Winkler Gasifier

From about 1929, Leuna implemented its operation with a completely new type of gasifier blasted with oxygen to manufacture water gas, similar to that of the slagging Wuerth gasifier, essentially absent of nitrogen. It was invented by Fritz Winkler (1888–1950), a chemist of BASF, to process the feedstocks such as coke breeze, coal fines, and low rank coals such as brown coals and lignites, which are typically difficult to process with the existing producers or the water gas generators. It is the earliest gasification operation adopting the principle of fluidization. To operate a Winkler gasifier the fuel bed of small fines or particles remains in a suspension in a bubbling mode like a boiling liquid by the draft of oxygen and steam blasted through a gas distributor. The Winkler gasifiers typically

Table 12.1 Syngas compositions of different gasification processes

	Wuerth slagging	Winkler	Pintsch-Hilbrand	Koppers-Spuelgas	Didier-Bubiag	Schmalfeldt
Feedstock	Brassert refuse	Brown coal	Brown coal briquettes	Brown coal briquettes	Brown coal briquettes	Brown coal as received
Oxidant	O_2/steam	O_2/steam	Steam	Steam	Steam	Steam/O_2
Heat source	Selfsustaining	Selfsustaining	Regenerators	Regenerators	External	Regenerators
Gas composition, vol%						
CO_2	9.7	19	14		9.4	18
CO	66.5	38	28	28.3	30.5	25
H_2	22.9	40	56	56.6	56.5	49.5
CH_4	–	2	1		1	3
N_2	0.9	1	1		2.2	3

Note Hydrogen sulfide and other hydrocarbon gases also present with some processes, not shown

operate at a temperature of about 900–1,000 °C, which is typically uniform throughout the bubbling bed. A good contact between carbon and oxygen/steam makes it possible to process a much higher throughput than the water gas generators or the producers. Some of the downsides though are the high sensible heat loss and the high carbon loss carried away with the resulting gas exiting the generator at almost the same temperature as inside the gasifier. To minimize the losses, a waste heat exchanger is placed right after the generator to recover the sensible heat to preheat the incoming oxygen and steam or an air while collecting carbon fines to recycle back to the generator.

The water gas or syngas produced with Winkler gasifiers has a high hydrogen content but relatively low concentrations of carbon monoxide and nitrogen and had therefore been deployed by four other hydrogenation plants developed a few years later; there were three Winkler gasifiers installed at each of the plants at Boelhen, Magdeburg, and Zeit owned by Brabag near Leipzig and five to six installed at Bruex (currently Czech Republic) owned by the Suddentenlandische Treibstoffwerke AG founded in October 1940. These Winkler gasifiers appear identical in design with a unit capacity of 20,000 m³/h of water gas. The Winkler gasifier, however, has its own limit as well; coals of highly active tend to perform well but some bituminous coals with a low reactivity do not. Although the Winkler gasifier is capable of processing a coke breeze and coal fines, already a step forward compared with all the previous sorts, it still has difficulty in processing small fines below 3mm because the small fines would be easily carried away without being reacted so that causing a high carbon loss.

Innovation moves on.

Creative Making of Syngas

The Tariff policy that the German government initiated in 1931 to levy import taxes on crude oil and oil products promoted the development of synthetic technologies. Such development was further enhanced when Hitler came into power two years later toward the industrial deployment targeted to increase the country's energy security by reducing its reliance on oil imports. By 1932, the majority of Germany's oil consumption relied on imports, primarily from the United States. At the time, Germany's oil consumption was still small relative to other countries, such as only about 8% of what the United States consumed and also far behind that of Russia and Britain. Such a situation remained unchanged until the mid-1930s with only a small quantity of liquid fuels produced from coal tars and the three demonstration plants at Ludwigshafen, Leuna owned by IG Farben, and at Oberhausen-Holten by Ruhrchemie AG. It had become obvious that both the synthetic technologies to turn coal into liquids, the hydrogenation and the Fischer–Tropsch, hold the golden key to a solution that would allow the country to turn its large coal reserves into the much needed liquid fuels. To execute such an industrial plan the

Nazi government in 1934 ordered the IG Farben to form a consortium, the Braukohle-Benzin AG ("Brabag"), with nine to ten coal companies to promote, develop, construct, and operate the coal to liquids plants. The government brokered the necessary license agreements with associated technology owners including IG Farben on its hydrogenation process and Ruhrchemie AG for its Fischer–Tropsch process (Stranges, 2003). By 1936 when Hitler announced the Four Year Plan, the Brabag through its subsidiary, the Gesellschaft für Mineralölbau GmbH, had placed three hydrogenation plants into services and one Fischer–Tropsch plant a year later. The number of synthetic plants then increased to seven hydrogenation plants and eight Fischer–Tropsch plants in 1939. By 1942, a total of twelve hydrogenation plants and nine Fischer–Tropsch plants were all in operation, of course with a few plants experiencing extended time of commissioning. The combined capacity of liquid products at the peak time during WWII in 1943 had reached about 4.5 million tons of liquid fuels including motor spirit, diesel, lubricant, and wax, representing about 69% of Germany's total consumption of liquid fuels. Remembering the fact that at the breakout of the war the combined annual production of liquid fuels from the two types of plants was only about 1.3 million tons; it grew to 1.9 million tons a year later. Such rapid growth continued each of the following years until the Allied strategic bombing hit in April 1944.

To visualize the gasification capacity that has to be built to produce the required syngas to support the synthetic operation, let's use the actual production numbers of liquid products to do a simple math. In 1943, the liquid fuel production reached peak with 3.4 million tons from the twelve hydrogenation plants and 0.43 million tons from the nine Fischer–Tropsch plants. Feeding such a large newly created synthetic industrial operation requires a significant amount of gasification capacity for syngas production, the carbon monoxide and hydrogen to the Fischer–Tropsch process and the hydrogen to the hydrogenation process. Based on actual performance at the time, each ton of liquid products would require about 8 tons of bituminous coal equivalent of which 6.1 tons would go to water gas making processes to generate the needed carbon monoxide and hydrogen for the Fischer–Tropsch operations. Assuming all the 0.43 million tons of liquid products in 1943 were made from bituminous coal it would require 2.6 million tons of bituminous coal to be gasified. For the hydrogenation operation, each ton of liquid products would require about 2 tons of bituminous coal equivalent to be gasified for the production of hydrogen and 1.9 tons of bituminous coal equivalent to be hydrogenated or directly liquefied. One ton of coal hydrogenated would need nearly the same amount of coal to be gasified so to produce the required hydrogen. Additional bituminous coal equivalent, 1.9 tons for the Fischer–Tropsch operation and 3.8 tons for the hydrogenation operation for each ton of liquid products, would be necessary to provide utilities such as heat and power required to maintain the operations. Obviously, the high operation pressure with the hydrogenation operations requires more energy for the compression of hydrogen and necessary streams. Producing 3.4 million of liquid products, assuming all from coal, would need to gasify about 6.8 million tons of bituminous coal equivalent to make enough hydrogen that would satisfy the hydrogenation operation. The combined use of bituminous coal for the two types of coal to liquids operations would amount to 9.4 million tons of coal to be gasified during 1943, an unprecedented demand for water gas or syngas.

Under such a circumstance, developing large gasifiers instead of building small ones would certainly be a logical thing to do. While working hard to develop larger gasifiers to provide such a large quantity of syngas for the emerging synthetic industry the syngas making had to deal with the challenges arising from the low rank coal feedstock. Germany has large reserve of low rank coals. A problem was that the most existing gasification processes are unable to handle such low rank coals. In doing so, coal gasification had to reinvent itself by demonstrating many creative ways to provide the needed syngas at a few of the synthetic operations. Here are some examples.

For those synthetic plants built in Western Germany during the early years of deployment, especially of the Fischer–Tropsch technology, syngas or water gas was largely made via conventional water gas generators, which is a non-continuous operation by alternating an air blast and a steam run. The gasifiers built into these plants such as the Steikholen-Bergewerk Rheinpreussen plant at Moers on the west bank of the Rheine, the Gewerkschaft Victor plant at Castrop-Rauxel, and the Ruhrchemie's Ruhrbenzin AG plant at Holten near Obenhausen were all fed with coke, which was produced from adjacent coke oven plants that carbonized coals to produce coal tars as the primary product. These synthetic complexes and many others developed at the time were typically vertically integrated with coking facilities and coal mining operations. Coal tars as feedstock were fed to the hydrogenation plants to produce liquid products, and coke was consumed by water gas generators to produce most of the required water gas or syngas for the Fischer–Tropsch operations and other hydrogenation plants, when applicable. The coke oven gas (COG), typically containing a high methane and hydrogen content, was steam cracked to supplement the water gas with hydrogen. A typical water gas contains hydrogen and carbon monoxide at a ratio of about 1:1.3 depending on coal, steam and operating conditions. To make the right ratio to 2:1 for the Fischer–Tropsch operations, additional hydrogen is required. At these early plants, it was achieved by supplementing hydrogen by a steam cracking of the available COG and with the balance made by a partial conversion of the water gas, typically less than a third, via the CO catalytic conversion understanding that the CO catalytic conversion was an expensive process.

In the first Fischer–Tropsch plant built at the Moers, for example, there were a total of 11 units of water gas generators equipped with waste heat boilers of the typical UGI type (shown in Fig. 7.3) but built at different times by the Koppers. The complex has a coke oven battery with 210 coke oven units from which about 980 tons of coke daily produced goes to the water gas generators and about 250,000 m^3 of COG daily to the steam cracking units. The steam cracking took place at 6 Cowper Stove steam cracking vessels lined with refractory and filled with checker bricks, and its operation is somehow similar to the regeneration operation. Under normal operation nine of water gas generators were in operation to produce 1.15 mm M3 of water gas daily. Like any typical cycle the

Table 12.2 Creative way to run water gas generators

Typical cycle	%	Modified with COG	%
Air blast	40	Air blast	38
Steam purge	3	Steam purge	2
Up run with steam	30	Down run with COG/steam	25
Down run with steam	25	Up run with steam	38
Purge run	3	Purge run	2
Repeat cycle		Repeat cycle	

operating cycle time of the generators was 202 s and allocated in the following manner (Table 12.2).

Water gas is collected during the runs. Then, about 18% of the water gas goes to the catalytic shift convertors to convert carbon monoxide to hydrogen by reacting with steam. The converted stream would be blended back to the water gas flow along with 390,000 M3 of the steam cracked COG. The combined syngas would have the right ratio of hydrogen to carbon monoxide. The combined syngas was then cleaned up to remove hydrogen sulfide by passing through processes, similarly to the coal gas making, and then organic sulfur before entering the Fischer–Tropsch converters. This setup for syngas making by shifting part of the water gas and supplemented with cracking COG for additional hydrogen, when available, had been deployed by many other Fischer–Tropsch operations including those mentioned above.

In another Fischer–Tropsch plant located at Bergkamen near Dortmund, however, the water gas generator had become more creative in its operation to produce a hydrogen rich syngas. The plant was developed by the Chemischewerk Essener Steinkohle AG in Essen, a partnership set up in 1937 by the Essner Steinkohlen Bergwerke AG and the Harpener Bergbau AG in Dortmund. The plant was similarly an integrated facility like many others with both coke and COG available for syngas making. Instead of steam cracking COG separately to make the supplementary hydrogen, the water gas generator integrated the COG cracking directly into its operation; the resulting syngas had a ratio of hydrogen to carbon monoxide at 2:1, well suited to the downstream synthesis. The plant was commissioned in 1939, becoming the 7th Fischer–Tropsch plant with a maximum annual capacity of producing 80,000 tons of gasoline and diesel products. Here is how the integrated syngas making works.

To execute such a scheme, the Chemische Werk Essener Steinkohle AG deployed a new type of water gas generators, the Demag gasifier, engineered by a local engineering firm, the Demag AG. The Demag gasifier appears essentially similar to the UGI or the G&H water gas generator, an operation by alternating air blast and steam run. But the tweaks made in the Demag gasifier was to inject a COG along with steam into one of the run cycle so that the COG gets cracked while passing through the incandescent bed of coke in addition to a water gas formation between the reaction of carbon and steam.

The deployed 12 Demag gasifiers consumed 800–850 tons of coke daily (unit capacity of 67–70 tpd) and 550,000 m^3/day of COG to produce 1.8 mm M3/day syngas for the Fischer–Tropsch synthesis. The blast and run cycle is 3 min, allocated uniquely in the manner as shown in Table 12.2.

Similarly, water gas is collected during the runs. An improvement was actually made to reverse the sequence between down run and up run while injecting the COG into the down run to make sure an effective COG cracking into hydrogen when the temperature in the coke bed was still high. Such a modification seems minor but very effective, both in cost and performance. The resulting syngas, containing 59% of hydrogen and 29% of carbon monoxide, was then water washed to remove the carried dust before entering a gas holder. After removing impurities such as hydrogen sulfide and organic sulfur the syngas would be ready to go to the downstream synthetic converters.

For the plants built in central Germany, however, more challenges arose and had become mostly unsurmountable for typical water gas generators when brown coal becomes the feedstock.

Special and Integrated Gasification

Over the past century, Germany has been the largest producer of low rank coals of lignite and brown coal in the world. In the mid-1930, Germany produced about 150 million tons annually, which peaked at about 430 million tons a year around the mid-1980s. Actually, most of the stakeholders in the Brabag consortium were owners of coal mining operations, which produced about 90% of the low rank coals in Germany at the time of forming the consortium. It appears a no brainer that any technology that could make use of the low rank coals to produce quality syngas for the synthetic processes would be of great interest, at least to these stakeholders.

During WWII, what the Germans did to support themselves with the much needed liquid products such as aviation spirit, motor spirit, diesel and lubricant etc. certainly impressed the Allied forces. Well before the war ended in November 1944, a committee was appointed by the Ministry of Fuel and Power to investigate and study the synthetic industry that Germany had developed to produce the war materials critical to sustaining the war. The committee chaired by Major K. Gordon of ICI was comprised of representatives of relevant companies such as ICI, the Gas Light & Coke Co., the Shell Refining and Marketing Co., Ltd., the Esso, the Anglo-Iranian Oil Co., Ltd., the Shell Petroleum Co., Ltd., and the British government agencies including the Ministry of Economic Warfare, the Ministry of Fuel and Power and the Fuel Research Station. The actual investigation also included the representatives from the US Bureau of Mines. One of the objectives of the investigation is to inspect the plants fresh as soon as those synthetic plants were uncovered or captured by the Allied forces. Some of the inspections actually took place within a few hours of its capture. Through such an investigation, a significant amount of information both technical and commercial about the synthetic plants and the operation of the plants had been collected (Power, 1947). Among the discoveries with regard to coal gasification, several special water gas making or gasification processes had caught the attention of the Allied forces. These special gasification processes such as the Pintsch-Hillebrand process, the Koppers-Spuelgas process, the Didier-Bubiag process, and the Schmalfeldt process are somehow similar to each other but also different in many details. On a high level, all of the processes share the following features, which were of particular interest to the Allied experts at the time.

- *Using lignite or brown coal as feedstock*
- *Continuous operation in water gas making*
- *Syngas manufactured with a steady composition*
- *Demonstrated in actual synthetic plant operations*

Then, the first three processes share a common feature at their core that the water gas generators are of essentially a vertical retort design, similar to the Woodall-Duckham vertical retort, where there exist three different zones, drying, carbonization, and gasification from top to bottom indicated as a, b and c in Fig. 12.1. During operation, coal is fed to the retort via the hopper at the top of the generator, moving downward through the three zones with the unreacted char or coke being discharged at the bottom, which is typically diverted to producers to manufacture fuel gas. To produce water gas, however, preheated hot steam, instead of air used in operating the Woodall-Duckham retort, is injected into the gasification zone to react with the red hot coke to produce the required water gas, which then goes up and passes through the carbonization zone and the drying zone. Part of the formed water gas is drawn at the middle of the carbonization zone to make the syngas for the Fischer–Tropsch synthesis after being cleaned up; the remaining water gas continues its path of moving up and driving out the volatile matters from the brown coal and leaving at the top of the retort as a recycle gas. The recycled gas would be cleaned up to remove dust and tars before being recycled back as feed or fuel. Here naturally comes the question of what available heating source was individually exploited to keep the temperature of the gasification zone high enough so that the endothermic reaction between carbon and steam would proceed to produce a water gas. In this regard, the Pintsch-Hillebrand process and the Koppers-Spuelgas process adopted the heat regeneration concept, invented by the Siemens brothers, to preheat steam along with the recycled hydrocarbon gas in one regenerator to about 1,100–1,300 °C before entering into the gasification zone while the other regenerator under heating by a producer gas blast (see Fig. 12.2 for details). Differently, the Didier-Bubiag process used producer gas along with the recycle gas to heat the retort externally via combustion, which is not shown here.

While the overall process for the Pintsch-Hillebrand process and the Koppers-Spuelgas process are essentially close, the major difference lies in the layout between the vertical

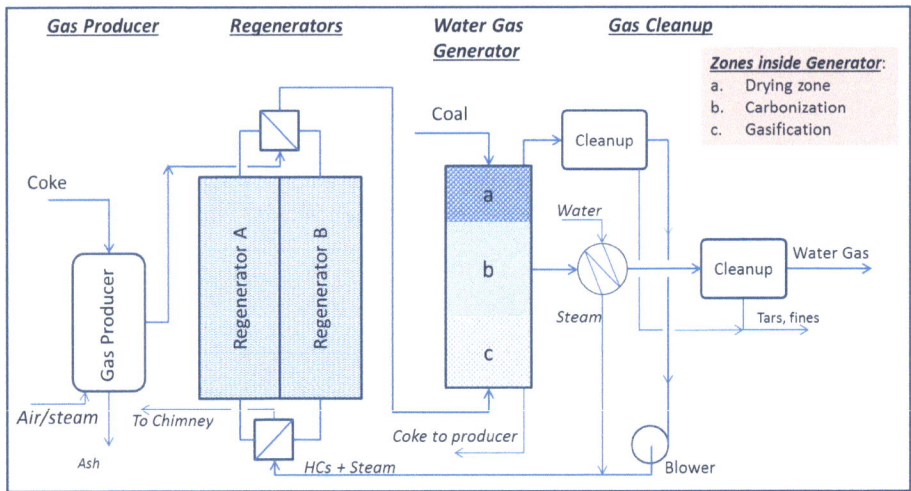

Fig. 12.1 Schematic process of the integrated water gas gasification

Fig. 12.2 The
Pintsch-Hillebrand water gas
generator (Power, 1947)

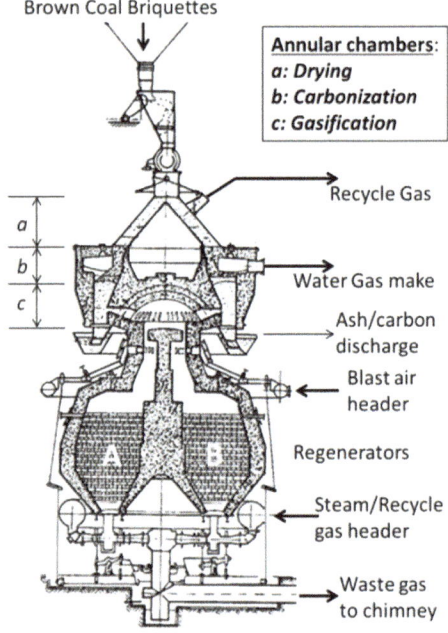

retort as the water gas generator and the regenerators. With the latter process, the regenerators are two sets of separate and stand-alone equipment from the vertical retort, which are situated on the same level as shown in Fig. 12.1. In addition, the designs of the regenerators are different as well but both are refractory lined vessels, which are filled with chequer bricks. Here are some details about the applications of the processes.

The Fischer–Tropsch plant, developed by the Brabag at Schwarzheide, north of Dresden, deployed both the Koppers-Spuelgas water gas process and the Didier-Bubiag water gas process to manufacture the needed water gas from local brown coal. It is the 4th Fischer–Tropsch plant that began its operation in 1937. The plant was originally designed with an annual capacity of producing 25,000–30,000 tons of gasoline and diesel and later expanded to 200,000 tons, becoming the largest Fischer–Tropsch plant in Germany. The local brown coal was made into briquettes, which were used as feedstock in the special water gas generators. During a normal operation, the brown coal briquettes are fed into the vertical retort via a coal hopper and pass down countercurrent the hot water gas rising up from the gasification zone and eventually become coke by losing its volatile matter. Part of the resulting water gas is drawn at the middle section of the carbonization zone to make water gas and the rest continues its journey up through the drying zone and leaves at the top of the vertical retort along with coal gas as the recycle gas. After removing tars and dust the recycle gas, joined by a stream of steam, circulates back to the regenerator that has already been preheated to 1,300–1,500 °C. The steam is generated with the sensible heat of the hot water gas in the waste heat boiler. While the recycle gas along with the steam becomes heated up by passing through the regenerator, the hydrocarbons contained in the recycle gas will be cracked as well by the high temperature along its passage up. At the exit of the regenerator, the recycle gas should be heated to about 1,100 °C before entering the gasification zone at the bottom of the vertical retort where water gas reactions and additional cracking of the hydrocarbon gases take place. The unreacted coke discharged will be consumed by producers to manufacture fuel gas that is used to heat up the regenerators alternatively.

The Koppers-Spuelgas generators, each unit with a capacity of 26,100 m^3/h of water gas, accounted for 80% of the required syngas for the Fischer–Tropsch operation. The remaining balance was provided by the Didier-Bubiag generators. After removing dust by water washing the combined water gas is further cleaned up with iron oxide to remove hydrogen sulfide and with sodium carbonate to remove organic sulfur. The final syngas ready for the Fischer–Tropsch synthesis has a ratio of hydrogen to carbon monoxide at 2:1 (Table 12.1). The cost of the produced syngas by the Koppers-Spuelgas process is much cheaper than that by the Didier-Bubiag process. There seem six units of the Didier-Bubiag water gas generators that were installed in the plant. The generators are essentially cylindrical tubes made of steel material. It did not use the regeneration as the heat source but used the recycle gas and the producer gas to heat up the carbonization zone through the steel walls in an annular passage surrounding the zone. More information can be found in the following references (Storch, 1945).

With regard to the Pintsch-Hillebrand water gas process, it is a technology that had been demonstrated at the Tiefstack Gas Works, Hamburg in 1932. The regenerator design is not stand-alone but actually integrated into the vertical retort as one piece of equipment (Fig. 12.2). By design, the vertical retort takes place in the annular space sitting right on the top of the structure of the regenerators. The recycle gas and steam preheated to 1,300 °C enter directly through a gas distributor slot into the bottom section of the gasification zone (*c*). Similarly to the Koppers-Spuelgas process the formed water gas moves up along the annular space and parts at the middle of the carbonization zone (*b*) as water gas make to exit and the remaining gas up to the drying zone (*a*), which then leaves at the top of the drying zone as a recycle gas. The regenerators are heated with producer gas alternately from top to bottom and the waste gas is discharged at the bottom of the generator.

The Union Rheinischen Braunkohle Kraftstoff AG (Union Rhenish Brown Coal Fuel AG) selected the Pintsch-Hillebrand water gas process to manufacture hydrogen for its hydrogenation plant. The plant was located on the west bank of the Rhine at the Wesseling, a village about half way between Cologne and Bonn. The Union Rheinischen Braunkohle Kraftstoff AG was established in 1937 to represent a group of lignite coal mine owners of the Rhine area to utilize its local lignite for synthetic projects. The construction work at the Wesseling started in 1938 with a capacity 250,000 of tons of liquid products a year. Due to the breakout of the war, however, the operation of the plant was delayed with its first liquid products produced in August 1941 and had since been able to increase its production steadily over the years until being air raided started in July 1944. By October 1944, the plant had been completely shut down due to the damage inflicted by the air raid. The plant was one of the early plants being inspected by the Allied Forces on March 18–19, 1945.

There were eleven units of the Pintsch-Hillebrand water gas generators and eighteen producers installed on site; both used lignite briquettes as feedstock. Each of the generators was designed to produce 5,400 m^3/h of water gas, which contains about 84% hydrogen and carbon monoxide combined as shown in Table 12.1. The overall water gas production would be total of 59,400 m^3/h should all eleven generators run at capacity at the same time. Additional hydrogen, if required, was supplemented by cracking the tail gas recycled from the hydrogenation operation. Then, additional processing such as cleanup, CO shifting and CO_2 removal follows, similarly to the ammonia process, to prepare hydrogen for the hydrogenation operation with exception that the hydrogen pressure was much higher, up to 700 bars. Producer gas was used as a heating source around the plant. Since the startup, the Pintsch-Hillebrand water gas process had been working well with a water gas rate maintained on an average at 48,600 m^3/h (82% of design capacity) during 1943 and 52,000 m^3/h (87.5% of design capacity) in 1944 until the time of the air raid (Europe, 1945).

To sum up, these technologies based on the principles of a vertical retort have been developed to continuously process low rank coal in the form of briquettes. By injecting a stream of preheated steam into the gasification zone of the retort, it essentially turns the retort into a water gas generator, which is capable of manufacturing syngas with a steady quality. By further integrating producers to digest the unreacted coke discharged from the water gas generators or coal briquettes as feedstock, the resulting producer gas would provide the needed fuel source to heat up the regenerators. These special and integrated water gas generating processes by combining the water gas making, the pyrolysis process and the producer effectively into a complex system did provide an efficient way to continuously generate a quality water gas to the synthetic plants at the time while utilizing the locally available low rank coals.

Solution to Gasifying Coal Fines

Gas making processes so far, no matter it is a gas producer, a water gas generator, the special water gas generator, and the like, has a common drawback; that is, these gasifiers are incapable of processing fines as a result grinding or a low rank coal such as lignite and brown coals which is typically very friable upon heating. Germany has significant reserves of lignite and brown coals which often contain moisture of more than 50%. Such a high level of moisture has to be driven off to a certain level, say below 14–15% by weight, by heating, either internally or externally. An unintended consequence of so doing is the fact that these coals tend to easily become pulverized, which is why these low rank coals need to be made into briquettes by using certain binders before being fed into, gasifiers, retorts or those integrated water gas generators previously discussed. In typical operations with a producer and a generator, the size of coal or coke needs to be in a certain range, typically between 8 and 50 mm. That is to say that the portion of coal below 8 mm would have to be disposed of by other means such as boilers or small furnaces. A gasifier that can handle fines below 8 mm would for sure be of interest. The Winkler gasifier was created partly for that reason, using coals of small sizes and coke breeze. For coals or coke breeze smaller than 3 mm including the pulverized low rank coals up on heating, however, the Winkler gasifier would have difficulty in processing them because of a significant carry over loss under normal operating conditions. For a floating or fluidized bed operation of coal, a wide range of particle size distribution tends to cause a high carryover of the small fines, which would result in poor performance due to carbon loss.

On March 9 and 11, 1945, a team of the Allied representatives made a site inspection at a synthetic plant located at Lutzkendorf, a brown coal mining district in central Germany. The team was made up of experts from private companies including the Gas Light & Coke Co., ICI, the British Fuel Research Station, and the US Bureau of Mines. There seems something unique about the plant, which otherwise would not attract these best outfits in the field of coal gasification and its downstream applications. The plant did indeed stand out in many ways from all other commercial plants associated with the synthetic industry. There are four units of a brand new gasifier installed and the units had been running with local brown coal *as received*. It is also a continuous and bulky process producing the syngas of quality for the Fischer–Tropsch synthetic operation. The gasification technology seems a unique and first of the kind by using powdered coal as feedstock, which had not been able to achieve steady operation until 1943. The plant since its startup had been struggling for years with both operation and performance. Then why did it draw so much attention?

The plant is the Mitteldeutsche Treibstoff plant, which was developed by the Wintershall AG, a salt mining company established in 1894, and was commissioned in late 1938 as the 5th plant of the Fischer–Tropsch type. The gas making process that the plant adopted is the Schmalfeldt gasification process, a first-of-the-kind technology invented by H. Schmalfeldt who was the plant director at the time. It was developed to handle a powdered coal or coal fines, quite an interesting and novel invention at the time. The local brown coal contains high moisture of 52% by weight and would easily become pulverized or powdered upon heating, just like many other low rank coals. Since its commissioning from late 1938, however, the plant appears to have been plagued with technical difficulties of all sorts and performed poorly most of the time until 1943 after which the plant operation had become steady; its operation had only reached a production of about 40% of its design capacity. Although the inspection team concluded in their report that "......*In its present state of development the process appears to be less economic than the alternative Winkler process, operating under similar circumstances. The producers are particularly inefficient. Because of its dependence on the availability of a very cheap and active fuel, the process appears to have little application to British conditions.*" it is nonetheless an interesting and inspiring process as the first of the kind that may well be worthy of a close look.

In general principle, the Schmalfeldt process is similar in terms of heat supply to processes previously discussed such as the Pintsch-Hillebrand process and the Koppers-Spuelgas, but a more sophisticated and highly integrated process. The integration covers not only a staged gasification system and a stand-alone regeneration set but also a coal drying system and a powered coal conveying system (Fig. 12.3). Saying it is the first of the kind really applies to a broader scope beyond the gasification system. On a high level, the brown coal as received is heated directly by the hot syngas coming from the gasification system and becomes disintegrated into powders, which is stored in a dry coal bunker after being separated from the syngas. The syngas, after removing dust, splits into syngas-make for export and a recycle gas. Then a part of the recycle gas goes to convey the powdered coal from the dry coal bunker into a staged gasification system; the remainder of the recycle gas is joined with steam and passes through the regenerators to shore up enough heat before entering the staged gasification system. The regenerators use a producer gas firing to stoke up enough heat with the chequer bricks filled inside for preheating. Here is a brief description of the overall process.

At the core of the plant, it is the Schmalfeldt gasification of a two stage design, a main generator and a secondary generator. Both generators are standing steel vessels

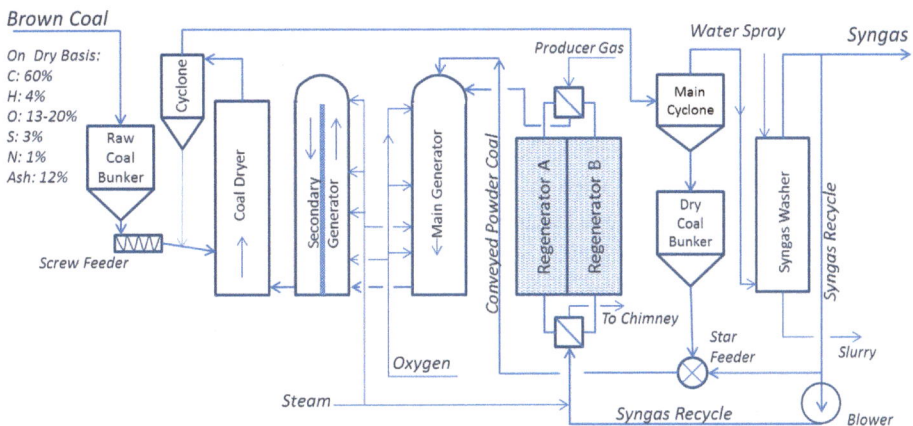

Fig. 12.3 Schematic process of the Schmalfeldt gasification

lined with refractory bricks, 5.5 m ID and 24 m tall, with a dome top and connected at the bottom. Unlike any gasifiers that had thus far been designed and practiced both the generators have no bed of coal, either a fixed bed like with a UGI generator and Wuerth slagging gasifier or a floating bed like with the Winkler gasifier, at all; they are simply empty vessels. During operation, the dried coal fine carried by a part of the recycle gas is blown through a small open passage at the center of the dome into the main generator where it meets the hot recycle gas and steam introduced through the ports on the side of the dome. Then gasification takes place to produce water gas. At the entry points to the main generator, the recycle gas loaded with the dried coal fine has built a pressure of 2.4 bars via a blower (not shown in Fig. 12.3) and the mixture of the recycle gas and steam has been preheated to 1,300 °C by passing through one of the regenerators. With the heat provided, gasification reactions should be able to sustain and maintain atmospheric pressure and 1,000 °C throughout the generators. With gasification going on, the flow of gases and the unreacted coal fines moves down the shaft of the main generator, exits at the bottom, and enters the secondary generator via a connecting pipe. The secondary generator has a partition wall to split the cylindrical vessel into two equal chambers. Then in the secondary generator, the flow moves up and then turns downward at the dome while additional gasification reactions continue. To maintain a constant and required temperature at 1,000 °C throughout the generator passage, oxygen or steam would be injected at different points along the vessel walls. Each of the generator units was designed to produce 25,000 m³/h of syngas. At the bottom of the secondary generator, the final syngas along with ash and unreacted carbon exits the second generator and enters the coal dryer.

The coal dryer is a brick lined cylindrical steel tower with 1.2 m ID and 20 m tall. The local brown coal once arriving on site is crushed and stored in the raw coal bunker from where the crushed coal is fed into the coal dryer via a screw feeder. The sloped entry point is located at a few meters above the bottom of the dryer. Once inside the dryer, coal of 52% moisture immediately gets engulfed in the hot rising syngas of almost 1,000 °C that would cause violent changes inside the coal, rapid evaporation of moisture and fast pyrolysis of the volatile matters. Quickly, the coal becomes dried and decrepitated into coal fines, which is easily carried away by the rising syngas. Now, the syngas, loaded with the dried coal fine, leaves the top of the coal dryer and then enters a cyclone to separate those large pieces, which is sent back to the dryer. The syngas loaded with coal fine along with ash resulted during gasification flows to the main cyclone where coal fines is separated and stored in the dry coal bunker. The syngas leaves the cyclone to the syngas washer with water spray to remove dust as sludge discharged at the bottom. The clean syngas leaves the washer at about 82 °C and then splits into 42% for export as syngas make and 58% as recycle gas. The syngas make has a combined hydrogen and carbon monoxide of 74.5% under normal operation, which would have a ratio of hydrogen to carbon monoxide at 2:1 (Table 12.1).

Now, the dried coal fines in the dried coal bunker contain about 60% carbon and 18% ash. A portion of the dried coal fine has to be removed from the system in order to maintain the ash content and sent to producers as feedstock to produce fuel gas.

With regard to the recycle gas, a small portion of about 10% of the syngas make after being provided with some pressure head (not shown in Fig. 12.3) passes through the star feeder to convey the needed quantity of the coal fine to the main generator. The majority of the recycle gas is boosted via a blower to enter the regenerator where it is joined with a certain amount of steam. The mixture of the recycle gas and steam is heated up while passing through the passage of the regenerator and leaves the regenerator at 1,300 °C before entering the main generator. In the meanwhile, the other regenerator is under producer gas firing to heat up the chequer bricks inside until the upper bricks reach 1,450 °C while the waste gas exits to the chimney at 450 °C.

In essence, the Mittledeutsche Treibstoff plant was a bulky, sophisticated, and highly integrated process that operated under a completely different principle that is that gasification takes place throughout the whole space within the generators in an entrained or suspension mode. This is a distinctive feature in comparison with a producer, a water gas generator, a Winkler gasifier or a Wuerth slagging gasifier. It appears not an ineffective approach to handle the coal fines. At the syngas make rate of 25,000 m^3/h, it requires about 29 tons of brown coal and a large volume of a recycle syngas, about 1.4 times the syngas make, plus a small amount of oxygen and steam to help maintain the temperature throughout the generators. Under such a condition, the residence time of coal fines entrained by the syngas passing through the generators in a very short period of time is estimated between 4.5 and 6 s. The Schmalfeldt process is essentially the early version of the entrained gasification technology dominantly deployed in today's market, but different in many ways such as that it is a non-slagging process, the required heat to maintain gasification is provided externally via regenerators and gasification takes place in multiple stages and so on.

The first of the kind Schmalfeldt gasification process was in some ways fully reflected not only in its poor performance (Table 12.1) but also in the extended duration of the startup, which had dragged on till 1943. In hindsight based on the Allied inspection report, the struggles experienced during the extended period of the startup attributed to all sorts and far beyond the gasification technology, ranging from design, engineering, workmanship, manufacturing to project management etc. A few examples are that the drainage system and the utility boiler were inadequately designed to handle certain upset operating conditions; that the lack of winterization left pipes and equipment open to frost bite during winter; that poor sparing consideration gave some equipment with too many while others none; that the Fischer–Tropsch downstream was unable to handle the high sulfur compounds due to the use of the high sulfur coal, and many more. Here are more taken from the report to help gain a glimpse into the plant startup.

> ......Moreover since the novel gasification processes required more than the usual amount of modifications in the first few months of running, they caused a heavy drain on what skilled labour there was, and so the factory could not be got out of the rut. There was a severe shortage of skilled engineers and foremen and the labour was mainly inexperienced, and as can be imagined discipline and morale were poor. In fact the whole reads as a classic example of how not to design, build and run a factory.

To help the startup change course, IG Farben in October 1939 dispatched about 187 experts comprising personnel of management, operation and maintenance to the Mittledeutsche Treibstoff plant site in an effort to trouble shooting and to resolve the difficulties that had been preventing the plant from a steady operation. This seems critical to help bring the operation out of chaos. During 1943, the plant produced a total of 330,774,000 m^3 of syngas, averaging a unit production of 12,600 m^3/h or 50.4% of the design capacity. In 1944 before the Allied strategic bombing, the plant reached 85.2% of its design capacity.

To recap, the strong demand for fertilizers and liquid fuels during the early part of the twentieth century opened up a totally different industry, the synthetic industry. The development of the catalytic conversion of carbon monoxide or the water gas shift reaction to convert carbon monoxide into hydrogen with steam, and the additional clean-up processes for the removal of carbon monoxide and carbon dioxide and other sulfur compounds had found a convenient application of the water gas generators. The subsequent incentives to utilize the largely available local low rank coals since the mid-1930s provided an impetus for coal gas making, the gasification process, to go through another round of extensive innovation after the last round, which was driven by the gas engine application started in 1879. New gasification technologies and processes were frequently developed to cope with different feedstocks. Examples are that the Wuerth slagging gasifier was deployed to destroy refuses or low reactive feedstock, that the vertical retort teamed up with the water gas generator to produce the syngas that met downstream synthesis, and that the regeneration process invented by the Siemens brothers back in the 1860s became once again a critical piece of equipment to transfer the most needed heat to the specially integrated gasification processes such as the Pintsch-Hillebrand process, the Koppers-Spuelgas process, and the Schmalfeldt process. The new development of the Winkler gasifier and the Schmalfeldt generators represent a significant step forward of pneumatic chemistry, the science of gases, where their properties and their working powers had been exploited to do works over the earlier centuries that also led to the establishment of chemistry before the end of the nineteenth century. The regeneration process, the Winkler gasifier, and the Schmalfeldt generator are certainly an extension in pneumatic chemistry in terms of utilizing gases to facilitate or enhance the gasification reactions in different ways. Of particular interest in retrospect is the last process, the Schmalfeldt generator, which incorporates the uses of gases not only to transfer heat but also to make the heat transfer seamlessly integrated with coal drying and coal gasification, respectively. By carrying coal fines into a refractory lined empty vessel, a multistage gasifier, where coal gasification takes place to its completion, here comes an effective solution for handling coal fines or low rank coals. This concept has completely changed the past gasification experiences, a retort, a producer, a water gas generator or an integrated one with some or all of them, always with a coal bed existent. Such an entrained gasification process in its early primitive form has laid the seeds for the modern entrained bed gasification processes to develop, grow and become largely deployed by industries about sixty years later.

References

Europe, U. N. (1945). *Tech Report No. 87-45, The Wesseling SynFuel Plant.* US Naval Technical Mission in Europe.

Holroyd, R. (1946). *Report on the investigation by fuels and lubricants team at the IG Farben AG Leuna works.* The Bureau of Mines.

Parker, J. R. (1923). *The nitrogen industry.* Constable & Company Ltd.

Power, M. O. (1947). *Report on the petroleum and synthetic oil industry of Germany.* The Ministry of Fuel and Power.

Storch, H. H. (1945). *Synthesis of hydrocarbons from water gas.* Wiley.

Stranges, A. N. (2003). Germany's synthetic fuel industry 1927–1945. In *AIChE 2003 spring national meeting.* AIChE.

Modern Gasification Process

13

The Energy market post WWII had dramatically changed, no longer driven by the at all cost war mentality. As a result, most of the synthetic plants in Germany had either been shut down or switched to other feedstocks other than coals except two plants being relocated to former Soviet Union and the Czech Republic, respectively, continued their operations. Coal gasification seems suddenly came to a halt almost overnight. What the IG Farben and the Ruhrchemie had achieved in the synthetic field prior to and during WWII, however, had certainly motivated many countries to look into the technological advances and to assess how the new technologies would work to benefit the needs of their own countries. As early as back in 1937, the National Research Council of the United States had realized the critical roles that chemistry played into the development of the synthetic technologies. The Council through its division of Chemistry and Chemical Technology set up a Committee on the Chemical Utilization of Coal to look into this matter and come back with recommendations to the Council on what needed to be done understanding the fact that the country has vast reserves of coals in addition to its dominant position in the production of crude oils worldwide. The Committee, chaired by H. H. Lowry, Director of the Coal Research Laboratory, Carnegie Institute of Technology in Pittsburg, had 40 member experts representing at least 20 companies, research institutions, universities, and government agencies etc. such as the Koppers Company of Pittsburg, the US Bureau of Mines, the Standard Oil Development Co. and so on. In January 1945, the Committee made its product available by publishing two volumes of thick books, the Chemistry of Coal Utilization, broadly covering the chemistry and knowledge that had been developed in the field of coals and its utilization; it is essentially a knowledge base of chemistry, technologies and its industrial deployment about coals and its utilizations. Upon further

Q. Zhuang, *From Coal to Hydrogen*, Synthesis Lectures on Chemical Engineering and Biochemical Engineering, https://doi.org/10.1007/978-3-031-55586-2_13

demand, the Committee was compelled to make two supplementary volumes, one published in 1963 and another in 1981, to keep the content fresh by adding new knowledge and technologies then developed. These volumes have become the "Bible" and well referenced utility books in the fields of both industrial applications and academic studies of coals and the utilization. In April 1944 before the war ended, the US Congress passed the *Synthetic Liquid Fuels Act* to authorize the Bureau of Mines to lead and develop a demonstration plant to demonstrate the hydrogenation process and technology. Bechtel Corporation was contracted to develop and built the plant. The plant, designed to process 200 barrels per day of liquid fuels, was in full operation in early 1949 on an old site of a synthetic ammonia plant in Louisiana, MO. Then with additional funding approved in 1948 under the same Act, the Koppers Co. was contracted to develop and build a plant of 80 barrels of liquid fuels per day by using the Fischer–Tropsch technology next to the hydrogenation plant, which started to produce oil in 1950 before reaching a full operation the year after. It appears that the demonstrations were unable to provide a clear picture that the synthetic liquid fuels would be competitive with the crude oil derived oil products in the US market, which is why the projects were short lived, being shut down around 1952. From gasification perspective, these demonstration plants might have used the typical water gas generators to generate the needed syngas or hydrogen because there was no other technology available at the time except some experimental pilot.

In general, activities around coal gasification post the war had remained low and quiet except for some small scale deployment for ammonia production. This was about to change when the government of South Africa decided to resume its 1935 plan to develop the Fischer–Tropsch project using its plenty indigenous coal.

The Lurgi Gasifier

Similar to Germany, South Africa has almost no exploitable domestic oil resources but a vast deposit of low quality coals that are subbituminous in nature and contain a high ash of above 30% (Table 13.1). The country had relied heavily on oil imports to meet its domestic oil demand. To counter such an increasing demand for foreign oil the government of South African in 1949 reissued the license to Anglovaal to proceed with the Fischer–Tropsch project that was interrupted by the war. Facing a changed financial condition, however, Anglovaal recommended the government form a state owned entity to take over the project. Based on this, the South African Coal, Oil and Gas Corporation was established in 1950, which is the current Sasol. The Sasol project is located on the south bank of the Vaal River, about 50 miles south of Johannesburg. Construction started in 1952 and finished three years later, which is the so-called Sasol I. Subsequent expansions led to the operation of Sasol II in 1981 and Sasol III in 1984. By that time, all three plants were churning out a total of 112,000 barrels of oil products daily and a large number of other chemicals (Anastai, 1980; Erasmus, 1987). With regard to technology selection, Sasol

Table 13.1 Coals adopted for Mark IV gasifiers at different commercial plants

Fuel type	Sasol I	Dakota lignite	Changzhi coal
M, %wt	10.7	34.3	0.3
Ash, wt%	32	9.5	20.8
FC, wt%	37.4	28.8	64.5
VM, wt%	19.9	27.4	14.4
Total	**100**	**100**	**100**
Gross heating value, kcal/kg	4,658	3,747	5,860
Ash melting point, T4, °C	>1,400		>1,500

made some changes to its original plan. For the Fischer–Tropsch synthesis Sasol decided to use a new fixed bed reactor, instead of the tube-plate design, developed by Ruhrchemie and Lurgi and another new circulating fluidized bed technology developed by a consortium of several US companies including Standard Oil Co., Hydrocarbon Research Inc. and M. W. Kellogg, which is currently KBR. To produce the needed syngas for the synthetic processes Sasol selected a pressurized gasification technology, the Luigi gasifier. What is the Luigi gasifier?

The Allied Mission did report the Lurgi gasifier in its report but labeled it not as of special interest as the Pintsch-Hillebrand generator, the Koppers–Spuelgas generator, the Didier-Bubiag generator, and the Schmalfeldt generator; the technology was categorized as of general interest to gas industry because it was used mostly to produce a high heating value fuel gas. The Lurgi gasifier is a technology developed by the Lurgi Gesellschaft fur Waermetechnik of Frankfurt in the early 1930s. It is the first continuous and complete gasification technology with oxygen and steam as oxidants that is operated under a pressure to produce fuel gas for domestic consumption because gasification under a pressure favors the formation of methane, the direction toward volume reduction (**Reaction 7 in Chap.** 10) and a pressure around 25 bar tends to optimize the methane content so to boost the heating value of the fuel gas. From the perspective of synthetic liquid fuels, however, methane has no value to the Fischer–Tropsch synthesis but a diluent; it is a loss of carbon with the gasifier that would result in lower efficiency. To manufacture the ideal syngas for the Fischer–Tropsch synthesis the Lurgi gasifier would need to drop its operating pressure to atmospheric pressure under which condition that the formation of methane would be minimized. Then, one may wonder why Sasol made such a choice for its synthetic operation where hydrogen and carbon monoxide were desired. While there must be many factors to consider the following quote may tell what was in Sasol's mind in its decision making back then (Erasmus, 1987).

> The coal ash melting property is also important. Sasol coal ash melts at about 1,400 °C. Since Lurgi gasification is a low-temperature operation, few operating adjustments are necessary to adapt to changing ash melting properties......The actual level of ash in the coal is also crucial.

The Lurgi gasifier is well suited to high-ash coals, since it relies to some extent on the formation of a nonslagged, dry ash bed at the bottom of the gasifier. High ash content is a negative factor in most other gasification processes.

From the experiences that many synthetic operations had gone through back in Germany, it seems that Sasol's decision in selecting the Lurgi gasification technology would hardly be an uneducated bet, but rather a well thought-through by giving adequate weight to the coal or coals to be gasified. In reality, the property and characteristics of coal are not only critical but in many cases dictate the selection of the appropriate gasification technology; a successful operation of coal gasification always requires that engineers design the gasification process around the coal feedstock so as to accommodate the specific properties and characteristics of the coal or coals as the feedstock. Otherwise, the road ahead would not be an easy one; there have been many lessons learnt till this day.

Similar to the UGI water gas generator or the Mond gas generator, the Lurgi gasifier is a cylindrical double-walled steel structure where the annular space between the walls runs with water to cool the inner wall during gasification. What is different is that the Lurgi gasifier is a pressurized system so that both its coal charging and ash discharging are designed with a locking mechanism to be air tight. It is a non-slagging oxygen gasification system moderated with steam. From an industrial viewpoint, the Lurgi gasification is a commercially proven technology. The earliest deployment at AG Sachsische Werke at Hirschfelde near Zittau was put into service in 1936. Two gasifiers of 1.15 m ID were deployed at the gasworks, each gasifier having a capacity rated at 625 m^3/hour of fuel gas. Based on the information provided to the Allied Mission by Lurgi AG and the AG Sachsische Werke, there were fifteen large Lurgi gasifiers deployed at two synthetic plants (one at Bohlen and the other at Bruex) to manufacture fuel gas between 1940 and 1943. These gasifiers performed well by gasifying brown coal lumps sized between 3–10 mm, maxed at 20 mm, under an operating pressure of 20 bar except that the heating value of the manufactured fuel gas reached only 3,900 kcal/m^3 (438 BTU/cubic feet), lower than the design number of 4,200 kcal/m^3 (470 BTU/cubic feet) at the Bohlen operation. The ten gasifiers installed at the Bohlen site actually took place at two separate times, the first five in 1940 and the next half in 1943. There were some design changes made to the five new gasifiers. Those changes include the coal charging hopper, grate driving mechanism and arrangements for scraping arms at the dome. Each of these gasifiers has an ID of 2.5m and was rated at 3,000 m^3/hour, much larger than the first one deployed in 1936.

The early Lurgi gasifiers delivered to the Sasol site are the largest at the time; each of the nine gasifiers is 12 feet ID (3.7 m) and rated at 23,100 m^3/hour of raw syngas. The operating pressure was designed high enough to feed purified syngas to the synthetic processes operated at pressures between 23 and 25 bars. It represents an improved design, known as Mark I, that Lurgi AG had developed during the post war time with Ruhrgas AG to process bituminous coals (LURGI DRY-ASH GASIFIER). Prior to this time, the Lurgi gasifiers had been working with low rank coals. Although the facility started up in 1955, Sasol could not avoid paying the tuition by going through a strenuous learning curve, pretty much like many previous coal gas making projects when facing the first of the kind technologies. It appears that Sasol was able to get over the many technical difficulties by making necessary changes, modifications and upgrading. The plant reached a satisfactory operation over about three years of time. In the meanwhile, such an exercise provided an opportunity for the Lurgi gasifier to leap forward to another level in terms of handling subbituminous coal and boosting its capacity.

Relative to the low rank coals, bituminous coals inclusive subbituminous coals generally have no issue with fines because they are geologically more developed and hold up well when being crushed, which favors the operation of the Lurgi gasifier or fixed bed gasification in general. Some of their features such as caking tendency and low reactivity, however, tend to create some operational difficulties for fixed bed gasification. This is why the fixed bed gasification including the Lurgi gasifier, water gas generators, and producers is not good at handling this type of feedstock. An example is the development of the Mond gas generator in handling bituminous coals. When Sasol commenced its gasifiers in 1955 the operation of the gasifiers appears only to last for a few months and the production of syngas reached only 80% of the design capacity. Changes and modifications had to be made to the existing gasifiers. It took about three years to increase the capacity to 27,500 m^3/hour of raw syngas, exceeding the original design capacity by 19%. Such an improvement led to the Mark II design. When the Mark III of the same inside diameter as the old ones, 3.7m, became available in 1966, the capacity of the Lurgi gasifier had increased to 34,000 m^3/hour of raw syngas, which is an impressive 47% jump from the Mark I design. Sasol added three Mark III into the existing fleet to replace some old ones. It is worthy of mentioning that there was a significant upgrade for the purification of syngas at the Sasol plant. Relative to the synthetic plants using water washing, Sasol projects adopted a new process for the purification of syngas, the Rectisol process, which uses chilled methanol to absorb all impurities including carbon dioxide, hydrogen sulfide, and an array of gum-forming compounds. It is a very effective cleanup process because methanol at a low temperature of -67 °F (-55 °C) has a strong selective absorption capability to these Impurities. The absorbed impurities can be recovered in a separate desorption step while the recovered methanol would be recycled back. The Rectisol is a process that has been used in many plants, as of today, where downstream synthesis catalysts have a stringent requirement for impurities typically carried over in a coal gasification setup.

Then ten years later when Lurgi and Ruhrgas developed the Mark IV, a slightly large gasifier of 3.8 m ID with a capacity of 46,000 m^3/hour, Sasol then added another three units in 1978 to replace the old units. By 1980, it is reported that the 13 gasifiers in services onsite had reached an average availability reaching over 84% (LURGI DRY-ASH GASIFIER) (Erasmus, 1987). For any gasification technology to achieve such a significant jump in coal throughput there must be a large design margin embedded or planned into the original design to counter any possible unknown risks associated with possible feedstock variances and unit operation. To release the design margin in the Sasol case, however, a good understanding of the subbituminous coal, its specific behavior inside the gasifier and the subsequent mass balance of different components are certainly critical factors that contributed to such a drastic improvement with both throughput and operation. Having developed the needed expertise in design and operation and the confidence with the associated technologies including the Lurgi gasifiers over the years, Sasol decided to move on to expand its coal to liquid fuels footprint in the 1970s. As a result, additional 40 Mark IVs were installed at the Sasol II plant located at Secunda, about 80 miles east of Johannesburg, and went into operation in 1980, and then a second set of 40 Mark IVs made up the Sasol III that went into operation in 1984. Ever since, the Sasol coal to liquids fuels initiative has morphed into a sprawling a coal derived petrochemical complex delivering several hundred thousand barrels of motor/aviation oils, 500 BTU fuel town gas, and 30 more chemicals to the local market (*the Sasol website*).

At the time, the success that Mark IV had achieved with the subbituminous coal in South Africa seems quite inspirational in a way that the Mark IV gasifiers would have a prosperous potential for more deployment. It is true somehow.

Then, the early 1970s witnessed the onset of the oil crisis, and in response to the crisis, Sasol was working hard to develop its Sasol II. Before the crisis was over, several natural gas companies in the Midwest region led by American Natural Resources formed a consortium, the Great Plain Gasification Associates, with the objective to develop and build a gasification facility to manufacture synthetic natural gas (SNG) by using the largely available lignite resources from a nearby coal mine in the Great Plains region. It appears that the project plan was to manufacture 250 mmscfd of SNG but reduced it to 150 mmscfd when construction began in 1981. Because of the commercial experience that Mark IV had established and its high natural gas content in the syngas produced from it, fourteen (14) Mark IVs with a design pressure at 427 psig (29 barg) were built into the Great Plains Synfuel Plant located at Beulah, North Dakota (Clusen, 1980). Different from Sasol and other previous coal gas making projects, what is waiting ahead for the project, unfortunately, is a totally different problem. When the project was completed in 1984, the energy crisis had been behind. According to the Bureau of Labor Statistics, the natural gas prices had dropped by 12% without an inflation adjustment, which would be 19% if adjusted from the peak time of 1981 when the construction of the project started and continued trending down. By 1986, the price had decreased by 31% without inflation adjustment and 45% if adjusted. Such a change in the natural gas market seems to have created an enormous financial pressure for the Great Plains Gasification Associates to continue operating the facility so that the US Department of Energy bailed out the company in 1986 and took over the ownership with $1 billion, about a half of the total price tag that the project was built. In 1988, another company, Basin Electric Power Cooperative who owns an adjacent power plant, acquired the coal to SNG facility under an agreement of profit sharing with the Department of Energy (NETL, Great Plains Synfuels Plant) (Co. D. G.). Since then the fourteen (14) Mark IVs had gone back to alive, consuming 16,000 tons of lignite daily to manufacture 150 mmscf SNG, which, after being compressed to 1,440 psig (99 barg), joins the interstate natural gas pipeline and is used for peaking purpose by the adjacent power plant. The Dakota Gasification Company becomes the operator of the facility. It appears that the Mark IV gasifiers worked well from the beginning of the operation with some minor changes to original design such as one of the changes made to the grate was to improve oxygen distribution. After all, the local lignite (Table 13.1) used at the plant is well within the experience range of the Lurgi gasifiers, which contributed to the fact that the actual operation has actually been more optimistic than design; for example 13 or all of the fourteen Mark IVs could be in normal operation instead of twelve operating with two in spare recommended in the original design (DOE, 2006). Similar to what the Sasol did, the company developed, eventually over the decades, more by-products such as tars, oils, fertilizer, phenol, and carbon dioxide for sale till this day.

Before 1990, the Lurgi gasifier has become the most deployed gasification process for large scale gasification operations because of its extensive experience with low rank coals and some non-caking or weak caking subbituminous coals worldwide. The technology seems ready to take another step to stretch its feedstock envelope. This time, it is an aged coal of a low volatile matter characterized between bituminous coal and anthracite.

In China, the Lurgi gasifiers had been deployed during the 1970s and 1980s. They were running relatively well with lignite at several plants such as in provinces of Heilongjiang and Yunnan. During the early 1980s, the Tianji Group (formerly the Shanxi Fertilizer Plant) located in Changzhi, Shanxi received the green light from the Ministry of Chemical Industry to build an ammonia plant with a capacity of 300,000 tons of ammonia annually by utilizing its local coal. Encouraged by the commercial success established by the Lurgi gasifiers, Tianji Group finally decided to move forward with the Mark IV gasifiers as well. Changzhi is located in the southern part of Shanxi on the north bank of the Yellow River. Like the whole Shanxi province, Changzhi is rich in coal resources but its coal differs from all the coals that Lurgi gasifiers had experienced thus far. The local coal features low moisture, high fixed carbon, and low volatile matter. It is a coal of high heating value compared to other coals that the Lurgi gasifiers had experienced (Table 13.1).

Changzhi coal is typically characterized as semi-anthracite or lean coal exhibiting a weak caking tendency. It also has a low reactivity, which was later found three to four times less reactive than the subbituminous coal used at Sasol I. Practically, it would be a reasonable fuel for UGI water gas generators because of its low volatile matter. The construction started in 1983, was completed about four years later in 1987, and then began quite a journey of learning. There are four Mark IVs built into the plant; each of them has an inside diameter of 3.8 m, a standard dimension of Mark IV, a water walled jacket structure, a mechanical grate, and scraping arms on the top of coal bed to break up any cakes formed from its caking tendency. Each of the Mark IV gasifiers was rated at 36,000 m^3/hour of syngas. During a normal operation, three Mark IVs would be running in parallel with one in spare so as to produce $3 \times 36,000$ m^3/hour that feeds the ammonia synthesis. From the onset of the startup, however, the plant had been plagued with all sorts of problems that lasted about eight years. Based on the design, a single Mark IV

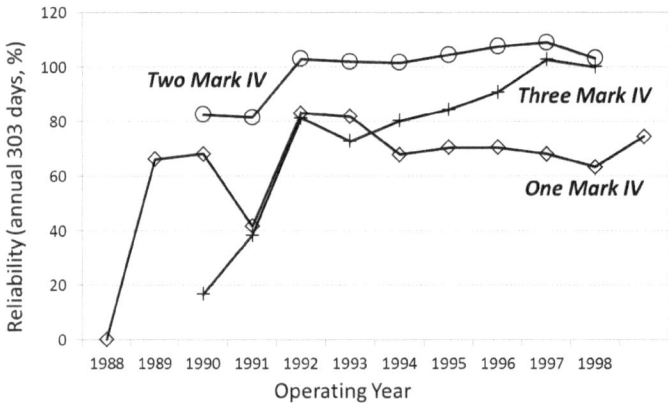

Fig. 13.1   Operation improvement at Tianji group over early years

should have a reliability of 75% and three of 100%. During the early years of operation, however, a single Mark IV gasifier shown in Fig. 13.1 was lower than the design target; two running Mark IVs were able to meet 100% reliability five years later after the startup and three operating Mark IVs had not able to meet the target until 1997 (Hongjian Jing, 2000).

Well, like many coal gasification projects, it is often difficult to single out one or two reasons that cause the operating problem but more often than not there is a range of issues that impact a normal gasification operation, especially when a coal gasification system had become a complicated process system. Some of the issues are technical and some are not. For Tianji's operation, for example, the quality control of coal was one of the major problems that appear to prevent a normal gasification operation. The Tianji Group received its coal from a local mining company that used a modern mechanical mining technology that produced fewer lump coals, which was worsened by the onsite inadequate sieving process so that the coal going to Mark IVs contained high coal fines, often higher than 10% in some worst scenarios. The design specifies a size range of 4–50 mm with fines below 5%. Such a large amount of fines created a serious carryover of the fines without being reacted to the downstream process where the fines and tars are mixed together during the subsequent cooling, causing plugging at the waste heat boiler and the tar separator. Then with more Mark IVs started up the quality of coal varies due to coal refuse or gangue content because coals were received from a multiple mining sources. These unpredictable variables impacted significantly the operation of not only gasification but also its downstream units including the sour catalytic shifting process and the Rectisol unit for syngas cleanup etc. Such situations would get worsened if there were no adequate design margins in these downstream processes so that the processes tend to have little wiggle room or flexibility to handle the upset situations. From the standpoint of gasification technology, however, there are also some lessons learned as well. The design of the Mark IVs provided to the Tianji project appears to have addressed the different nature of the Changzhi coal with mechanical scraping arms in place to counter the potential caking of the coal and by predicting a higher exit temperature of raw syngas at 650–700 °C, which is normally in the range of 265–400 °C for low rank coals. In retrospect, the design "upgrades" may still underestimate the impacts of such a high temperature profile on the operation of the Mark IV gasifier. The Changzhi coal, which featured as high heating value and low reactivity, contributes to an increased combustion zone, which would subsequently create several undesired situations, a high unconverted carbon, and a high temperature through zones of gasification, pyrolysis and drying. The high uncovered carbon would cause not only a low thermal efficiency but also a high temperature of ash discharging, resulting in serious damage to the mechanical ash grate and the lock hopper valves right down below. The high temperature profile within the Mark IV gasifier contributed to damages to the water jacket and the mechanical scraping arms as well. These damages caused a significant amount of downtime in the plant operation in addition to the capital spent on equipment, parts and repairs. In 1996, Tianji Group had some technical exchanges with Sasol about the operation of Mark IV and learned the impact of coal quality as well to the operation of Mark IV at Sasol. Tianji also made a purchase of several lock hoppers for ash discharging from Sasol, which reported a significant improvement in the ash discharging experience. Later that year, Tianji decided to switch to washed coal instead of mined coal in order to maintain the needed consistency of coal quality. Then, after a few more important design changes made to tar recycle, grate, water jacket water circulation etc. the plant was finally able to run three Mark IVs at the same time and achieved 100% reliability in 1997.

To a certain extent, the unfortunately prolonged learning curve at the Tianji plant stems from a combination of the lack of knowledge of the coal, its behavior at a high temperature, and inadequate design consideration to counter any potential upsets from the operation of the Mark IVs. Identifying such root causes early on and making necessary remedies, however, is always a challenging task considering such a complex system and the many variables impacting a gasification operation. From a technology perspective, what Tianji Group has achieved to get over the learning curve has helped the Lurgi gasifier make another step forward, after what the Sasol did thirty years ago, to further expand its hands-on experience on feedstock by adding another data point. An inconvenient truth about coal gasification is that the estimation of the impact of coal on a gasification operation still largely depends on hands-on experience. In later years, Tianjin Group added more Mark IVs to expand its annual ammonia production to 450,000 tons. In 2010, Tianjin Group merged into the Lu'An Chemical Group Co., Ltd. of Shanxi. At about the same time, additional six Lurgi gasifiers, an upgraded version, were built into a coal to liquid fuels plant located in Lu'An, owned and operated by the company. Syngas produced is shifted and purified before reaching a Fischer–Tropsch synthesis process, a technology

developed by a local company adopting the principle of a slurry bed operation, to produce an annual 160,000 tons of liquid fuels. The project became one of the first two coal
to liquids projects commercialized on an industrial scale around 2010 in China, about a
half century later after the Sasol I and 73 years after the first Fischer–Tropsch project, the
Steikholen-Bergewerk Rheinpreussen plant at Moers-Meerbeck, Germany with an annual
capacity of 70,000 tons.

It is generally agreed upon even in today's market that the Lurgi gasification remains
one of the few technologies suited for handling low rank coals of high moisture and coals
of high ash and high ash melting temperature. In the past ten years, there have been half
a dozen of coal to SNG projects being placed into commercial operations in China; these
projects are being executed in phases. There are a large number of Lurgi type gasifiers,
somewhere between 80 and 90, currently in operation. The number would be more than
doubled if these projects are fully implemented. Low rank coals or subbituminous coals
are feedstocks going to these plants.

The Search for Modern Gasification

Technologically speaking, what Germany had achieved across the two wars in opening
up a new synthetic industry represented a significant milestone for science, technology,
engineering, and industrial operation, which their interactions could produce miracles
like what Frederick Woehler was excited in 1828 about his discovery that human beings
would no longer need a living kidney to make urea. With the breakthrough achieved by
BASF, urea can be produced amply from the unlimited supply of ammonia synthesized
from hydrogen and nitrogen. This has unleashed the potential capability of agriculture to
produce enough crops and food to feed the rapidly growing population worldwide ever
since; that chemicals like methanol can be artificially made instead of distilling wood
chips; and that liquid fuels can be synthesized, instead of relying on mother nature, to
keep machines, automobiles, and airplanes running day and night.

Intriguingly enough, the core behind the miracles is the fact that all these artificial or
synthetic processes need are two small molecules, H_2 and CO, which thus far had been
produced via the coal gasification. By reflecting on how coal gas was made to feed those
synthetic processes, however, it is obvious to notice that skills, knowledge, and know-
how that are necessary with gasification in order to provide the quality and quantity of
syngas, ultimately the two molecules of H_2 and CO. For example, the large number of
gasifiers was deployed to manufacture enough syngas to produce 30 tons of ammonia
daily at the Oppau plant, and the hundreds of coke ovens had to be built to provide
the needed feedstock for the production of water gas. At Leuna, the Winkler gasifiers
were developed to digest the coal fines left from coal crushing, and the Wuerth gasifiers
were created to consume the unconverted carbon that had to be pushed out from other
water gas generators. To tackle these challenges presented with low rank coals, some

of the coal to liquid fuels plants developed several complex and sophisticated special water gas generator processes, all featuring an incomplete gasification, to consume the local low rank coals to manufacture the most needed syngas for downstream synthetic processes, Fischer–Tropsch or hydrogenation, and more. On the one hand, there seems no lack of innovation and ingenuity in so doing. On the other hand, these generator systems were expensive to build, and costive and cumbersome to operate as with the cases of the Pintsch-Hillebrand generator, the Koppers–Spuelgas generator process, and the Didier-Bubiag generator process. Due to the incomplete gasification with steam, the unconverted carbon along with ash after being discharged had to be utilized by producers to manufacture the fuel gas that is used to fire up regenerators, boilers and other onsite applications as utility. When the war was over the driver of a *do at all cost* mentality no longer existed, and so were these processes.

With the energy market changing post the war, the demand for gasification had scaled back dramatically because crude oil and natural gas had returned back to the commercial market. For example, the two oil pipelines, the Big Inch and the Little Inch developed in 1942 and 1943, to transport oil from the Southwest to the Northeast were privatized in 1947 and then converted into natural gas pipelines, which had essentially marked the end of the age of water gas in the United States. As a result, the number of gasifiers in commercial operation dropped to about 2,000 in 1948 from over 11,000 in 1921 (Corp, 1980). In Germany, gasification operations stopped overnight except for some ongoing development works. Globally, small ammonia and methanol plants continued to be built here and there where water gas generators were deployed to produce syngas. Overall, the post war efforts on gasification technology somehow accented to the development of an effective complete and continuous gasification process, that is, to gasify coal to its completion, for the production of syngas but not fuel gas as much as possible. The mature gas producers of all types at the time certainly falls out of this category but they are used to manufacture fuel gas, where it makes sense. Funded under the *Synthetic Liquid Fuels Act*, the United States was now taking the lead in developing a better gasification technology for the production of syngas in a strategic way. Under the Act in 1946, the Bureau of Mines launched an extensive program at its Morgantown Station to investigate coal gas making process aiming at demystifying the gasification of coals in order to develop a better cost effective gasification technology (NETL, History).

In principle, the Lurgi gasifier reflected such a trend as a complete gasification process developed post war by Lurgi AG, which explains why it was soon deployed by the Sasol I in South Africa and some other countries as one of the few available gasification technologies during the 1950s. In the meantime, the Lurgi gasifier is not an ideal technology from the viewpoint of the synthetic industry because its syngas contains a significant amount of methane and certain quantities of tars, ammonia liquor and phenol, just like many existing coal gas making processes, which require additional process units to recovery and purify the wanted syngas for downstream synthesis. In markets where these by-products are valued and valued highly, the by-products would be able to help improve the bottom line of the synthetic operation, which otherwise would be a drag, making the project less favorable. Obviously, South Africa goes with the former scenario, which is why the Sasol I project took its root by not only selling liquid fuels but also selling a 500 BTU/cubic feet fuel gas by blending methane with a certain amount of off gas from the Fischer–Tropsch operation, via a pipeline developed in later years to the local market. Sasol also makes sales of additional dozens of more by-products valuable to the market. A few examples are that tar is distilled to obtain creosotes, road pitches and light naphtha, which naphtha is hydrotreated to make benzoles as a solvent or a blending stock for gasoline. Phenols are extracted out of the ammonia liquor and used to make more organic products and ammonia is then stripped with steam to make fertilizer, ammonia sulfide, and more. What Dry and Erasmus of the Sasol Technology (pty) Ltd. presented in *"Ann. Rev. Energy"* in 1987 sounds more insightful.

> The production of oil from coal is viable in South Africa because of the unique combination of three factors: (a) the absence of petroleum deposits, (b) the availability of low-cost coal near the main market (the Johannesburg area), and (c) the high cost of transporting oil to Johannesburg, which is 400 miles from the coast at an altitude of 6,000 ft. ……Because of by-products produced in Lurgi gasification, additional recovery and work-up plants are required. The by-products, however, make a positive economic contribution. The inability to use by-products, say due to small scale operation, will obviously be a negative factor in a comparison study.

Considering that each ton of the Sasol coal gasified gives off 6 U.S. gallons of tar and oil and 19 lbs. of ammonia recovered plus the 10% methane gas in the raw syngas, it appears important that any similar project should make sure that there would be a market for these by-products. So is true for the half a dozen of coal to SNG projects executed in the past ten years in China. Then in the United States, the Dakota coal to SNG project has a different story of success to share. The owner of the Great Plains Gasification plant has found a market for its carbon dioxide, a concentrated stream already separated from syngas thanks to the fracking development with the oil and gas activity in North America. The owner has been able to sell the concentrated carbon dioxide via its 200 miles pipeline to Saskatchewan, Canada for enhanced oil recovery since October 2000, which helps improve its bottom line. After all, these by-products are part of the process of coal gas making from day one and had impacted the coal gas business in both ways, which is dependent on times or markets. Fundamentally, the formation of the by-products is inherent to the gasification processes such as their specific designs, the temperature distributions inside gasifiers, and operating conditions under which the coal gas making takes place. These processes include the Lurgi gasifier, the gas producer and the water gas generator as well. With regard to the operating conditions, all these processes thus far had been operated at atmospheric pressure and under ash melting point. Although the combustion takes place at temperatures below the ash melting temperature in any of these gasifiers, the countercurrent flow of the rising gases and the down flow of the cold coal from the top would create a wide range of temperature profile, which is the primary cause of the by-products. In addition to the formation of the by-products, the fixed bed or moving bed designs themselves inadvertently created another bottleneck that would make the scale up difficult.

Entering into the age of an industrial synthesis, however, the quality requirement for a coal gas has shifted from being a fuel gas with a high heating value to a syngas of as rich hydrogen and carbon monoxide as possible. A gasification technology that could break down the by-products into hydrogen and carbon monoxide would of course be desired. In the meanwhile a high capacity gasifier would reduce the number of gasifiers so to save the costs of equipment, operation and maintenance for the production of syngas, which typically makes up the majority of the costs to build and operate a synthetic plant. Here, the fixed bed design of gasification including the Wuerth slagging gasifier has its inherent barriers that are impossible to circumvent.

Under the principle of fixed bed gasification, the capacity of a gasifier or the rate of coal reacting with oxygen and water vapor is essentially dictated by how fast the reactions take place within the zone of combustion, typically a layer of $4''$ to $6''$ depth sitting on top of the grate. Once coal or a mixture of coals and the final product for the project are decided, the required gasification conditions should accordingly be optimized. What now determines the capacity of a gasifier would become the cross area of the zone of combustion, the inside diameter of the shaft area. It is common sense that a gasifier with a larger diameter tends to have a higher capacity for processing coal under the same operating conditions. To a certain extent, therefore, a unit capacity of a gasifier becomes a parameter that represents the level of not only the gasification technology but also the capabilities of engineering and equipment manufacture. Then, the intensity of gasification defined as the unit capacity on a unit cross area of the shaft of combustion zone actually reflects the level of the gasification technology. What is shown in Fig. 13.2 are the changes in the unit capacity and the intensity of different fixed bed gasifiers including the Wuerth gasifier and the Winkler gasifiers against the shaft diameter of each individual gasifier in the order over a time span from 1890 to 1980. So the horizontal axis is not in scale but in the order of the gasifiers deployed commercially. The Winkler gasifier is provided here as a comparison though it is not a fixed bed gasifier. The unit capacity is expressed as coal rate per unit gasifier, tons per day, and the intensity as the unit capacity per unit of shaft cross area, tons per day per square meter. It is obvious that, except for the three Winkler data points, the unit capacity in general goes in hand in hand with its corresponding intensity; that is to say that, building a fixed bed gasifier of a large unit capacity requires a larger gasifier. Although each data point represents an operation running with different coals so that the comparison between data points may not be exactly apple to apple, the trending of the intensity data point should reflect the level of technology advancement. The intensity trending suggests that the Lurgi I gasifier has made a big step forward compared with the Mond gas producer or the water gas generator but during a time span of more than half a century. Then a slow uptrend follows from the Lurgi I to the Lurgi II, the Lurgi III and the Lurgi IV. Although the Lurgi V is a much larger gasifier of 4.7 m ID

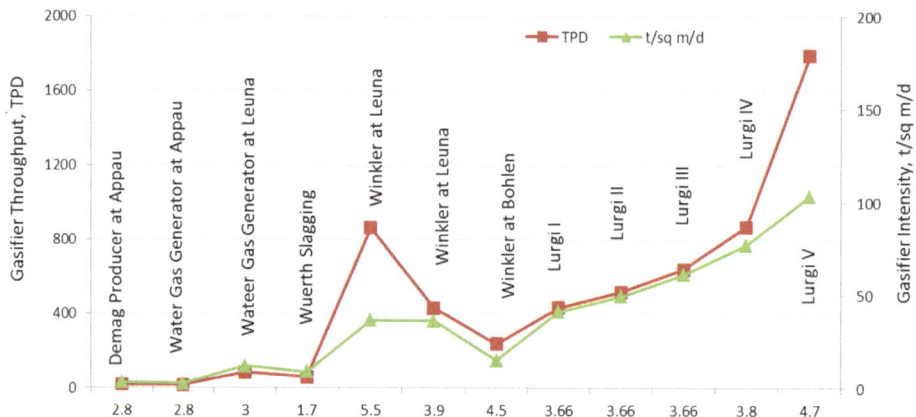

Fig. 13.2 Evolution of capacity and intensity of fixed bed gasifiers (Note that the horizontal represents inside diameter of gasifier at shaft area, not in scale)

that was developed at Sasol around 1980 there appears no further industrial deployment. The forty gasifiers at the Sasol III, developed in the early 1980s, also adopted the Lurgi IV. From an industrial perspective, a gasifier of 3.8 m ID was the largest deployed until entering 2000. Going larger than that might have faced challenges, which could be related to operations, feedstock, equipment manufacturing, and logistics and so on, which are difficult to overcome back then.

Although under a fluidization principle the Winkler gasifier somehow still resembles a fixed bed. It is different from a fixed bed gasifier that there is no distinction between different zones of combustion, gasification, pyrolysis and drying, and the reactions of coal with oxygen and steam take place, theoretically, quite uniformly throughout the whole bed at the same temperature. This explains the high intensity of the Winkler gasifiers relative to the previous Mond gas producer and the water gas generator. It seems a technology that has potential for further development, or at least, deserves another look. With regard to the Wuerth gasifier, it shares the same principle of a fixed bed gasifier but operates at a higher temperature at the combustion zone so that ash is discharged as a slag. Although it is about 40 +% smaller relative to the Mond gas producer and the water gas generator in terms of inside shaft diameter its intensity of gasification is on the same level, representing a technological advance. When coal is the feedstock, however, the Wuerth gasifier would not be able to prevent the formation of the by-products as with any other fixed bed gasifier.

Moving forward, it appears that a gasification technology that not only delivers a high intensity of coal gasification but also produces a high quality of syngas with minimum by-products would be the direction to go to benefit a modern synthetic industry. How to achieve such an objective and would it be possible to achieve it? The commercial experience that had acquired with the Winkler gasifier, the Wuerth slagging gasifier, the special Schmalfeldt water gas generator and the Lurgi gasifier might have suggested that the necessary parameters and principles to engineering modern gasifiers focus on the previous operating experiences such as suspension gasifier or fluidized bed, coal fines or coal particles or lumps, slagging operation or non-slagging, and pressurized system or ambient etc. The only parameter of a certainty at the moment would be that gasification should adopt oxygen and steam as the oxidants, instead of air and steam. Others would have to be weighed out.

Then, the challenge comes down to how to piece them together and integrate them into a working product, the modern gasifier.

Entrained Gasification

The post war activities on gasification development had clearly shifted to the principle of conveying coal fines into a gasifier as being practiced at the Mitteldeutsche Treib-stoff plant at Lutzkendorf, the special Schmalfeldt gasifier process because coal fines had been recognized for having the following benefits, (1) speed up gasification reaction with oxygen and steam due to its small fines of typically less than 500 μm; and (2) broaden the range of coals including caking coals due to the intact nature of the fines suspended inside a gasifier. The Schmalfeldt gasifier was actually further explored by the French after the war, and a pilot plant, west of Paris near Rouen, was built in 1950. Similar to the Schmalfeldt gasifier, it is a cylindrical refractory lined vessel, capable of processing about 1,600 lbs. of coal per hour. Coal fines were conveyed with air axially into the gasifier at the top center of the gasifier dome. Oxygen and steam, preheated to a high temperature with a regenerator, were introduced coaxially around the coal feed line. It is called the Panindco process and the tests were conducted with coal containing a high volatile matter under a non-slagging operating temperature. Another process called the Bianchi process was trialed as well at the Marseille Gas Works under the similar condi-tions as the Panindco process except for that the gasifier was tested at pressures up to 25 bars (Elliott, 1963). Similarly in the US and Germany, most development activities with regard to gasifier design and operation started to focus on the following design features by moving to high temperatures.

- Entrained bed operations adopting coal fines as feedstock
- Oxygen and steam as oxidants
- Slagging operating temperatures

What was still diverging is the operating pressure, some at ambient pressure and others at pressure. This last parameter somewhat becomes what would eventually define the modern gasification technology.

Ambient Pressure Entrained Gasification

There was a gasification process that fits some of the above criteria but went under the radar screen of the Allied Mission inspection at the end of the war. It may be that the technology never got deployed in any of the synthetic plants. The gasification technology is known as the Koppers–Totzek gasifier producing syngas, free of nitrogen, tars and oils, by injecting coal fines with oxygen and steam into the gasifier under temperatures high enough that coal ash is discharged in a molten slag mode. The technology was an ambient pressure system invented as early as in 1938 by H. Koppers Co. of Essen, which is now the Krupp–Koppers Co. GmbH. There appears cooperation between the Koppers Co. of Essen and the US Bureau of Mines around 1950, which the latter developed a pilot plant processing 24 TPD of coal at Louisiana, Missouri. The first commercial plant was offered to a company to produce syngas from gasifying 50–100 TPD of a bituminous coal in France in 1949. The next is an order received from Typpi Oy Company, Finland in 1950 to produce 60 TPD of ammonia from a Polish coal at its works at Oulu. The plant, designed and constructed by Koppers Co., went into operation in 1952 (Dierschke, 1955). Since then about a dozen of plants worldwide had adopted the Koppers–Totzek gasifiers to meet their ongoing demand of ammonia fertilizers by the 1970s.

The Koppers–Totzek gasifier is a horizontal cylindrical vessel, either lined with refractory or a water jacket walled design (Fig. 13.3). On both ends opposing to each other are where coal fines are blown into the gasifier chamber via screw feeders along with oxygen and steam. Subjecting to a radiantly hot environment inside the gasifier, typically at a range of 1,500 and 1,600 °C (2,700–2,900 F) above the ash melting temperature, the coal fines immediately react with oxygen and steam to its completion, much faster than any previous gasifiers. The resulting hot raw syngas leaves at the top and enters right into a waste heat exchanger where steam is raised. During operation, part of the molten ash flows down to the bottom of the gasifier and is discharged into a slag sump and the rest is carried away as dust by the hot raw syngas and gets separated at the waste heat exchanger upon cooling on its way up. Ash carryover from the Koppers–Totzek is generally high and could be significant depending on the coal and operating conditions. Koppers claimed at the time that its gasifier could be designed to process a wide range of coals from lignite, subbituminous, bituminous including caking ones, anthracite and even petroleum coke. Because of the high temperature inside the gasifier, the resulting raw syngas contains essentially no detectable tars and oils, which have been cracked down to small molecules. Methane in the syngas is only about 1,000 ppm.

As an example, there is an ammonia plant built at Modderfontein, north east corner of Johannesburg, South Africa. Commenced in 1974, the plant was designed to manufacture 1,000 tons of ammonia per day by deploying six Koppers–Totzek gasifiers of a two headed design to produce syngas for hydrogen production. Compared with the hydrogen making process engineered at Appau, the Koppers–Totzek gasifier based ammonia process had become much simplified due to the improved quality of syngas (Fig. 13.3).

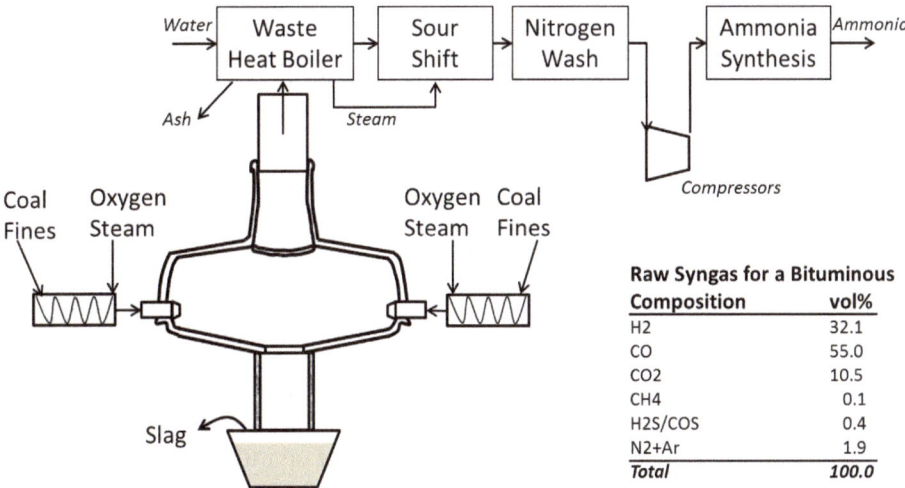

The table in the figure:

Raw Syngas for a Bituminous	
Composition	vol%
H2	32.1
CO	55.0
CO2	10.5
CH4	0.1
H2S/COS	0.4
N2+Ar	1.9
Total	*100.0*

Fig. 13.3 Sketch of Koppers–Totzek gasifier for ammonia production

A few years later, Koppers Co. started to offer a larger gasifier, which is a four headed design, 90 degree to each other. Such a design had been deployed at two ammonia plants in India. Each of the plants was designed to produce 900 tons of ammonia daily and both of them were placed into operation in 1979. The new gasifier with a four headed design is about double the capacity of the two headed design that was offered to the Modderfoutein plant in South Africa. The four headed Koppers–Totzek gasifier has a dimension of L7.9 m (26ft) × W7.9 m and H7.805 m (26ft), each gasifier rated at 35,000 m^3/h (1.24 million cubic feet/hour) of raw syngas, typically containing 86% of H$_2$ + CO. The Koppers–Totzek gasifiers are quite bulky.

Pressure Entrained Gasification

During the 1950s, several other entrained gasifiers operated under an ambient pressure were developed as well such as the Babcock-Wilcox du Pond gasifier and the Rummel single shaft gasifier. The former was developed jointly by the US Bureau of Mines and Babcock & Wilcox at the Morgantown station of the Bureau of Mines and a full scale gasifier was later installed into one of the du Pont Company's plant at Belle, West Virginia in 1955. The Rummel single shaft gasifier was a water-walled cylindrical vessel developed by the Union Rheinische Braunkohlen Kraftstoff AG at Wesseling. Different from the Koppers–Totzek horizontal design, these two gasifier designs adopted a vertical cylindrical vessel design, which should be easy to fabricate, maintain, and occupy a smaller floor plan. However, they all share the same principles as an entrained up-flow gasifier with oxygen and steam as oxidants, and are operated under slagging temperatures and ambient pressure. With the Babcock & Wilcox du Pont gasifier, a refractory lined reactor is divided into a lower chamber and an upper one where coal fines carried by a stream of hot oxygen and steam are injected into the lower chamber at multiple locations around the chamber. Subjected to the radiant hot environment inside the lower chamber, the coal fines react immediately with oxygen and steam to its completion. Hot syngas rises up into the upper chamber where it is cooled and ash becomes slag, which falls down to the slag sump. Here, there is a common feature with all entrained bed gasifier; it is the fact that the exit temperature of the syngas is almost the same as the reaction temperature, which is why there is a need to install a waste heat exchanger immediately downstream of the gasifier to recover the sensible heat from the red hot raw syngas so to reduce the potential heat loss. To accommodate the materials of the waste heat exchanger or boiler, however, the raw syngas needs to be cooled down while passing through the upper chamber to about 800–900 °C prior to entering the waste heat exchanger and leaves the waste heat exchanger at a temperature equivalent or lower than that of a fixed bed gasifier. Similar design consideration was applied to the Rummel single shaft gasifier as well except for that the Rummel single shaft gasifier is a one chamber design without division as did with the Babcock & Wilcox du Pont gasifier. Such a waste heat exchanger has become a critical piece of equipment for most entrained gasifiers. Also, there is a pool of slag remaining at the bottom of the Rummel single shaft gasifier where multiple nozzles are used to inject coal fines. Differentiating from previous entrained gasifiers, the Rummel single shaft gasifier uses recycled syngas to convey coal fine into the gasifier through the nozzles, which are arranged between injection points for oxygen and steam, which is an improved design than that to convey coal fines with oxygen and steam, often causing explosion due to potential flashback of flow. A full scale plant was developed at Wesseling to produce 560,000 scf of syngas per hour containing 84% hydrogen and carbon monoxide (Elliott, 1963; Stroud, 1981).

So far it appears that an entrained gasifier operated under a slagging temperature could deliver a quality syngas with no or minimum formation of tars and oils, and could also expand the feedstock portfolio by using a wide range of coals. The question now remains how an operating pressure would impact the gasification. The Bureau of Mines also investigated this perspective up to 31 barg in a refractory lined 10″ ID gasifier of 20 cubic feet at its Morgantown Station. So did the Institute of Gas Technology (IGT, currently GTI) in a smaller pilot unit at a pressure of up to 7 bars. Both of these pressurized gasifiers were downflow, instead of up-flow, refractory lined cylindrical vessels. The investigations at the Morgantown Station focused on several critical design related issues such as methods of introducing reactants into the gasifier, reactor designs, and the impact of pressure on gasification capacity etc. It tried to inject the reactants (coal, oxygen and steam) tangentially into the upper part of the gasifier via multiple burners, but soon gave up due to a high refractory erosion rate. It finally landed on using a pressurized fluidized bed feeder to inject the combined reactants axially at the top center into the gasifier as did with the Schmalfeldt gasifier. The data points of gasification at different pressures suggest an almost linear relationship between the operating pressure and the syngas production capacity, which supports the development of a pressurized entrained gasifier. It is interesting that the IGT's gasifier design separated the introduction of oxygen from coal fines, where the oxygen was introduced as a down jet axially from the top of the gasifier while coal fines were conveyed by superheated steam tangentially into the upper cylindrical part through several locations around the gasifier. The IGT investigations used bituminous coals of both caking and non-caking.

From the development and testing experience of these gasifiers, it is not difficult to draw a simple conclusion, that is, the development of gasification technology at that time was on the right track. In addition to the design principle of a pulverized coal entrained flow reactor operated above the ash melting point, the gasifier operated under pressurized conditions has obvious advantages, and seems to reflect the characteristics required for a modern coal gasifier. The challenge, however, is how to feed the pulverized coal into a pressurized gasifier in a reliable and safe manner. Luckily enough, among all the entrained gasifier processes that were under development at the time, however, there was a process that stood out, not only accommodating all the beneficial parameters and principles discussed above but also possessing some unique design features that make the process resilient and would become one of the major processes in a widespread industrial deployment a few decades later. It is the Texaco gasification process.

The Texaco gasification process is an entrained down-flow gasifier operated under conditions of slagging temperatures and at pressures, similar in many ways to the Koppers–Totzek gasifier, the Babcock & Wilcox du Pont gasifier, and the IGT's as well. It is a technology developed by Texaco Inc. an oil company established in Beaumont, Texas in 1902. It appears that Texaco started the development of a coal gasifier as early as 1948 with a pilot facility of about 15 tons of coal daily. Based on the tests of about two dozen of solid feedstocks, a demo unit was developed in 1956 at a plant of Olin Mathieson Chemical Corp. in Morgantown, West Virginia. The demo gasifier was designed to operate at a pressure of up to 34 bars and to process about 70 tons of coal daily to produce 3.5 million cubic feet of syngas for the production of ammonia (Elliott, 1963).

What makes the Texaco gasifier unique is that it blends coal fines with water to make coal slurry so that the slurry could be pumped into the gasifier conveniently and safely. The Texaco gasifier is a refractory lined cylindrical vessel, divided into two parts with the upper part as the refractory line gasification chamber and the lower part as a pool of water called the quench chamber. Between the two chambers is a specially designed narrow passage. In its early design, coal slurry was pumped to pass a heater to raise its temperature high enough so that water inside would evaporate once reaching a cyclone located on top of the gasifier. A portion of the steam carries coal down into the gasifier where it meets with oxygen injected separately and the reactions takes place immediately. The resulting syngas along with molten ash moves downward and exits the gasification chamber through the narrow passage into the quench chamber where syngas disengage from the molten ash and leaves the gasifier at the side of the quench chamber. The molten ash drops into a pool of water and becomes solidified before being discharged at the bottom of the gasifier. It appears that keeping coal fines and oxygen separate has been one of its design principles in coal fines handling, which is reflected in one of its early patents filed in 1948 (Moor, 1953). The later development by injecting coal slurry directly via an injector into the gasifier with oxygen through an annular space around the injector has become part of Texaco's proprietary information package, which is still in use today. Decades later, the use of a coal water slurry feed method handling pulverized coal was an important step in the commercial success that the Texaco coal gasification process would have achieved.

The idea of making solid coal into a liquid form (slurry) and injecting it directly into a gasifier is indeed an ingenuous invention, which completely avoids the hazardous situation created by using oxidants of oxygen and steam to convey the solid coal. The slurry injection makes the design and operation of a pressurized gasifier convenient with the aid of pumps of diaphragm type. Although it appears that water evaporation inside the gasifier would sacrifice, to a certain degree, the cold gas efficiency the overall benefits by using coal slurry, nevertheless, would outweigh the potential downsides including the loss of cold gas efficiency. In some cases, on the contrary, the coal slurry feed would be one of the ideal processes because the high concentration of hydrogen in the syngas would reduce the burden of downstream CO shifting should hydrogen or hydrogen-rich syngas be the desired products. This would especially make more sense in today's market toward building a hydrogen economy. Well, how did Texaco land on such a simple and almost perfect idea? There must be a logical process for the decision. An obvious explanation, on one hand, could well be that it simply comes down from the DNA of Texaco Inc. as an oil company from day one; it becomes used to handling liquids. On the other hand, coal slurry is not a new thing at all by the 1950s. The Londoners actually built and ran a slurry pipeline in 1914 to transport coals from the Thames River docks to their city. Slurry transportation had also been practiced by mineral mining operators for a long time. The post war economic activity in the US created some tensions between railway operators and coal mine owners, which led to the development of coal slurry pipelines to distribute coal from mine mouths to markets far away during the decades from 1950. It was championed by the Consolidated Coal Company and led by the American engineer and inventor, Edward J. Wasp (1923–2015) who joined the company at Pittsburg in 1950. Wasp conducted an extensive fundamental investigation and conducted loop tests to understand the rheology and characteristics in relation to designing and engineering the pipelines. His work led to the construction of the 108 miles (172 km) Ohio coal pipeline, which was commenced in 1957. Wasp joined Bechtel in 1963 the same year when the pipeline was shut down and soon made Bechtel a global leader in the pipeline business for the purpose of moving all sorts of solid/liquid products. Several more coal slurry pipelines developed during 1960–70 are, for example, the 273 mile Black Mesa pipeline from Arizona to Nevada in 1970 and the development of the longest coal slurry pipeline running from Wyoming to the lower Mississippi river during 1970s (Partridge, 1982;

Whipple, 2014) (1981). By 1970s, the principle of coal slurry and relevant engineering experience had been established.

The coal slurry fed Texaco gasifier process developed during the 1970s had become a prototype of its future development, as indicated in one of the patents granted in 1970 (Slater, 1970). The subsequent industrial scale demonstrations, first at Oberhausen in late 1970s and then at Cool Water, California between 1984 and 1989 had acquired necessary data and experience for future industrial deployment. In these demonstrations, coal slurry was directly injected into gasifiers.

By the way, what should not be ignored and played an important role in Texaco's success in coal gasification process is the fact that the development of the Texaco coal slurry gasifier process has certainly benefited from its commercial experience with the partial oxidation ("POX") process of gaseous and liquid feedstock that started from the late 1950s. The post war economic recovery in Europe was boosted by the discovery of oil and gas both in the North Sea and in the Middle East during 1950–60. The increasing demand for hydrogen and carbon monoxide for the production of oxo-chemicals and for refining hydrotreating had created an opportunity for deploying the POX to manufacture syngas, which is typically more economical compared with the approach of coal gasification. Texaco seems well positioned itself to offer its POX process to the market in the 1950s. Its early licenses include two POX contracts using natural gas as feedstock to produce syngas, one with a chemical plant in Ravenna, Italy in 1958 and the other one with the ICI the following year. The POX process built for ICI was to replace the old water gas generators for the production of ammonia at the Billingham site. This is the year when the first large Koppers–Totzek gasifier was put into industrial service to produce 270 tons of ammonia daily in Greece with more followed entering into the 1960s in the fertilizer market. The demand for POX in the petrochemical sector, in the meanwhile, picked up as well, and Texaco delivered two more POX licenses to Mitsubishi Chemicals Co. and Daicel Chemicals in Japan, respectively. Both plants went into service in 1961. Not long after another Texaco licensed POX plant was placed into operation at Gela, Italy, another player joined the POX market by offering its own POX process. It is Royal Dutch Shell Oil Co. (Shell) with its first POX license that went into industrial service in Finland in 1965. Ever since the POX market to produce syngas from natural gas, refinery off-gas or liquid hydrocarbons including the heavy residuals from the bottom of the barrels have been dominated by the two players, Texaco and Shell, till this day (Fig. 13.4). Although the Koppers Co. or its successor claimed that its Koppers–Totzek gasifier could also handle liquid feedstocks there was no report on its actual industrial deployment outside the ammonia market. To sum up, the two players delivered a total of 11 POX projects during the 1960s and another 17 projects in the 1970s. The market cooled down a little bit in the next decade and came back up with a peak of 22 POX projects during the 1990s. The active deployment of the entrained coal gasification projects did not happen until about early 1980s.

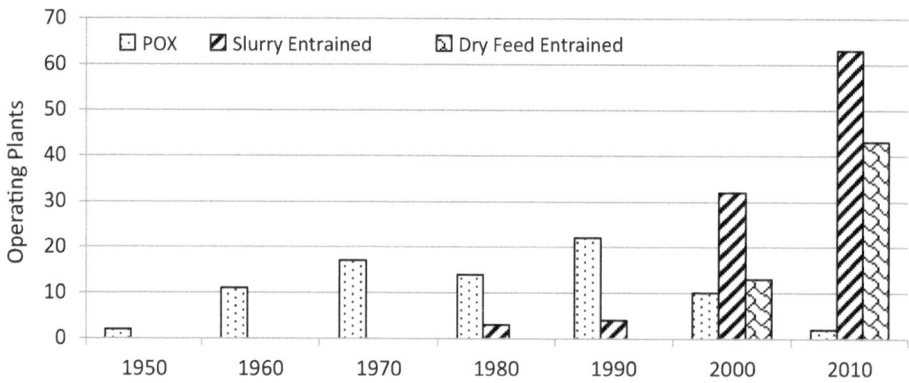

Fig. 13.4 Global POX and pressure entrained coal gasification deployment during each decade (data covers up to 2017 and excludes projects for power generation)

The disruptions in the oil market in the 1970s created another wave of interest in the development of technologies and related processes from coal to liquids. To provide a better syngas to it the pressure entrained gasification process also received its attention as well in order to tap into the vast coal reserves in the US. With government and industries investing multi hundreds of millions of dollars, many more companies had joined the development rush from which a dozen of additional coal gasification technologies were born in addition to the Koppers–Totzek gasifier and the Texaco entrained gasification process. They are the E-Gas gasifier developed by Dow Chemicals (currently owned by Lummus Technology), the Noell gasifier by a company in Germany (called the GSP, currently owned by Siemens) (Burgt, 2003). In Japan, an ambitious initiative named the Sunshine Project was launched by the agency of Industrial Science and Technology in 1974, aiming to develop alternative energy sources to oil. Two gasification processes were under development by Mitsubishi and Hitachi, respectively, as well (AO Technology, 1981).

Texaco's POX gasifier is a cylindrical refractory lined vessel and a downflow system with gaseous or liquid feedstock and oxidants compressed or pumped via an injecting device into the gasifier axially from the top of it. Different from the Texaco entrained gasifier, the Texaco POX gasifier is typically coupled downstream with a syngas cooler. Once coming out of the POX gasifier, the raw syngas along with undecomposed steam would go through the syngas cooler to pass its heat for a steam generation while being cooled down. Because of the better quality of either a gaseous or a liquid feedstock, which also contains little mineral matter, the POX system is a relatively simple system compared to an entrained coal gasification process. Once converted to syngas, however, the subsequent cleanup and purification would be essentially the same regardless of feedstock types from the viewpoint of a process. Such experiences that Texaco had acquired from its POX

licensing practice must have helped Texaco in the development of the entrained coal gasification process, not only from gasification perspective but also from system engineering
and facility operation and maintenance. Furthermore, its hands on experience as an operator in the petrochemical industry would be valuable as well. This mindset in turn might
not have helped Texaco when it comes down to marketing as Texaco seems not paying attention to the development outside its domain focus such as the fertilizer market.
Instead, Texaco stayed focused on syngas demands with the petrochemical sector from
the many licenses sold from that point on. Around 1980, Texaco wrapped up the demonstration of the entrained coal gasification process at Oberhausen. The pilot gasifier had a
capacity of 160 tons of coal per day and was operated at a pressure of up to 80 bars. The
pilot operation provided necessary data and operating experiences important for future
industrial applications (Corp, 1983). Also in 1980, funded by Tennessee Valley Authority
("TVA"), US EPA and EPRI, a field test with Illinois #6 coal was conducted with the
Texaco coal gasification facility at Oberhausen. The primary objective was to evaluate
leaching characteristics and toxicity of slag and trace metals in effluents around the Texaco coal gasification facility. The results would be used to help TVA evaluate its planned
IGCC project at the time. A syngas composition was reported as shown in Table 4.1 V in
Chap. 4, clearly having a high hydrogen concentration.

Among all other gasification technologies developed during the time of the oil crisis,
most of them faded away afterward. There is one technology that stood out and achieved
success in the long run, competing with Texaco's entrained coal gasification in the market
place. Again, it is Shell who developed the Shell coal gasification process. Not like what
Texaco Co. did, Shell took a different business approach to enter into the technology
space of coal gasification by partnering with someone who had a solid track record with
coal gasification. Shell selected the Koppers Co. who became the Krupp–Koppers GmbH
with the objective to develop a highly efficient entrained gasification system that would
be operated under pressures and slagging temperatures. The joint development kicked
off in 1974 and had gone through a 6 TPD scale lab facility commenced at the end of
1976 at Shell's lab in Amsterdam and then a 150 TPD pilot plant built in 1978 by the
Krupp–Koppers at a refinery at Hamburg owned and operated by Deutsche Shell AG,
a subsidiary of Shell (Burgt, 1981). A typical syngas composition is listed in Table 4.1
VIII in Chap. 4, featuring a high concentration of carbon monoxide because of the use of
dry coal feedstock.

The Shell–Koppers gasifier appears to take a similar design as the Koppers–Totzek
gasifier did in terms of diametrically arranged injectors but later evolved into the current
form as a straight down cylindrical, water walled, and up-flow gasifier. The water wall
design is similar to the membrane walls used in conventional pulverized coal boilers.
There are four feed injectors oriented horizontally in the mid-section of the cylindrical
gasifier. Deviating from the Koppers–Totzek design with a waste heat exchanger sitting
right on top of the gasifier, the Shell design relocated the waste heat exchanger as a
separate down-flow unit. In its later design, syngas resulted from the fast reactions of

coal fines with oxygen and steam flows up the gasifier and slag goes down to a water pool at the bottom of the gasifier before being discharged through a lock hopper system. The up flowing syngas gets quenched directly by a recycled cool syngas to about 900 °C. The combined flow then turns 180 degrees through a transition pipe and enters the waste heat exchanger, also known as syngas cooler, and then a dry solids filter to free syngas from the entrained dusts. In 1981, however, both parties agreed to terminate their joint efforts and instead to pursue separately their own business interests. As a result, Krupp-Koppers developed its own coal gasification technology, a dry-feed, membrane walled entrained gasifier operated under pressure and slagging conditions named the PRENFLO, currently, the Uhde process, which was later deployed at the 250 MW IGCC project in Puertollano Spain under the auspices of the European Commission in the late 1980s. Shell continued its path by establishing the Shell coal gasification process (SCGP), which was lined up with several projects for electricity generation in the Netherlands and Germany during the 1980s.

From the beginning, the Royal Dutch Shell appears to gear toward developing a high efficient coal gasification technology that would best suit for the generation of electricity. The company had already been working on several prospect projects that were lined up in that direction, which is publically announced by the representatives of the company in 1981 (Burgt, 1981).

> The next step will be the construction and operation of one or two 1,000 t/day prototype plants (*Author: IGCC plants, one in the Netherlands and the other in Germany*) which are scheduled for commissioning in 1983-4. Towards the end of the 1980s large commercial units with a capacity of 2,500 t/day are contemplated (*Author: which became the 250 MW Demkolec IGCC built in 1994 in Buggenum, The Netherlands*). The economy, especially of these large size units, is very competitive.

To accommodate the purpose of generating electricity under the principle of an IGCC concept, the Shell gasification design adopted a dry feed injection system and a syngas cooler, which is typically more efficient than a coal slurry gasification. Of course, the SCGP could also be applied to other purposes where hydrogen and/or carbon monoxide are required, which would be subject to a decision based on project economics. However, designing a pressurized gasifier integrated with a dry coal fine had been a challenge not only back then such as with the special Schmalfeldt generator and the Koppers–Totzek gasifier but also today. This is especially true when it comes to producing chemicals and liquid fuels as the syngas should be free from an inert gas such as nitrogen. Shell's design had selected a lock hopper system to achieve the objective, a bit more complicated than what had been practiced by the many water gas generators, producers and Lurgi gasifiers. To feed coal into a pressurized gasifier the coal must first be dried and finely ground in a mill where warm, inert gas flows through the mill to facilitate the removal of the coal moisture. The dried coal is then pressurized via a lock hopper system where several hoppers work in a lockstep sequence to transfer coal fines from one hopper to the other into the gasifier without causing any interruption of the pressurized gasification operation (Burgt, 2003). The SCGP gasifiers operate at pressures of up to approximately 45 bars. Under such a pressure, the resulting syngas after subsequent cleanups would have enough head pressure before entering a gas turbine of a combined cycle island. No additional syngas compression is normally required. Except the Demkolec IGCC project, unfortunately, none of other IGCC projects were materialized because of the fact that the oil crisis went away early 1980s.

Deployment of the Pressure Entrained Gasifiers

With the driving force for coal to liquid fuels disappearing, governments and some associated industries and companies in the developed countries had turned their focus on what Shell had contemplated, to develop IGCC projects, aiming to develop the second generation technology for power generation, which would be a huge market if successful. Such an environment seems to mirror what Mond was excited about when he placed his hands on the fuelcell technology about a century back for electricity generation. The IGCC in principle remains the same as what Mond did back then except for the facts that gas engines and dynamos have been upgraded to turbine technologies. Texaco and Energy Power Research Institute ("EPRI") took an initiative to launch an IGCC demonstration project to substantiate such a development at Cool Water California, which was later joined by GE with its 7 frame E class gas turbine, Bechtel with its EPC expertise, and the US Department of Energy under the Clean Coal Technology Program (NETL website). The demonstration project used bituminous coal to produce syngas by using Texaco gasifiers of a quench design and a syngas cooler design to feed the combined cycle to generate 96 MW net during 1984 and 1989. The syngas composition manufactured at the Cool Water shown in Table 4.1 VI & VII in Chap. 4 appears marginally better than the pilot test with the same coal at the Oberhausen plant back in 1980, implying the 1,000 TPD demonstration scale at the Cool Water could perform better than the 160 TPD scale at Oberhausen. The acquired information and experience have laid the foundation for a series of IGCC projects both in the next decade and the decades after. Here is a list of them (Table 13.2).

The confidence in Texaco gasifier undoubtedly stems from its commercial experience. Before the start of the Cool Water demonstration project the Texaco entrained-flow gasifiers had been used commercially at three industrial sites. In Tennessee, three quench gasifiers began operation at a pressure of 65 bars in 1983, using bituminous coal to produce syngas that is fed into the production of methanol; four quench gasifiers are operated at 37 bars to produce hydrogen for the manufacture of ammonia in Japan; and an 800 TPD gasifier with a syngas cooler design was due to start operation at the time in Germany for chemical production (Alpert, 1986). From the handful of the IGCC projects that have materialized during the past thirty years (Table 13.2), however, the road to market for the

IGCC technology must have been a strenuous and bumpy one. Like many coal gasification projects dealing with the first of the kind, each of the IGCC projects has its own feature and, therefore, has gone through its own learning curve, some lessons learned are related to feedstock, and more are to gasification, gas turbine, engineering, construction, and operation etc. So far, the IGCC itself has not been able to prove its competitiveness in the market place, facing either the natural gas fed combined cycle technology or the ultra-supercritical pulverized coal boiler technology for the generation of electricity. As a matter of fact, the gasification islands at several plants have already been either mothballed or shutdown, for example, the Elcogas IGCC, the Demkolec IGCC, the Tampa Electric IGCC, and the Kemper IGCC. As a complicated system as it can possibly be, the IGCC, taunted as the second generation technology for power generation, would have to exercise more patience in its long journey down the road in order to achieve the much wanted success. Nonetheless, the demonstrated excellent environmental benefits such as the low emissions of NOx, SOx, PM2.5 and the flexibility and capability to capture CO_2 shall prove that IGCC would have a play in the ever challenging market of the hotly debated energy transition. So far from the perspective of industrial deployment, the synthetic industry has become the bright spot for the pressure entrained gasification process, almost 80 years later.

The early 1980s witnessed three licenses for Texaco entrained coal gasification process, two in Japan for the production of carbon monoxide and ammonia and one at Kingsport, Tennessee for the production of acetic acid. Entering the 1990s, the market of coal gasification started to shift to China market, and four plants using Texaco's pressure entrained gasifiers went into service in China market. Except for one plant to manufacture town gas, the other three are to produce syngas for ammonia synthesis. Due to the fast developing economic activities in China, the decade after the millennium brought up an explosive growth for the modern gasification processes as a whole, 45 gasification plants had been placed into commercial operation between 2000 and 2009, among which 71% are slurry fed type of modern gasification processes and the rest are dry feed type of modern gasification processes including the Shell's SCGP. Shell started to market its gasification technology to serve the market around 2005. Such a momentum continues and there are about 106 plants went into commercial operation during the seven years from 2010. Although most of the plants are up and running in China, however, there are a couple of changes in the gasification landscape. First of all, the demand for coal gasification has emerged in other countries such as Korea and India. Secondly, more technology players have joined the force such as the E-Gas, the GSP, and a hand full of local technologies developed in China in addition to Texaco's and Shell's, which both of the coal gasification processes are currently under the house of the Air Products and Chemicals Inc., headquartered in Allentown, Pennsylvania. These technologies are all of modern entrained gasification in principle, either slurry feed or dry feed, and each with its own specific design variants. Thirdly, the growth rate in China for the slurry gasification slows down a bit to 59% while the dry feed gasification makes up the rest, a big step up. It

Table 13.2 Commercial IGCC projects worldwide (most information cited from the NETL website)

Project/Location	Gasifier	Gas turbine	MW (net)	COD
Demkolec IGCC/Buggenum, the Netherlands	Shell gasifier	Siemens V94.2	253	1994
Wabash River IGCC/Indiana, USA	E-Gas gasifier	GE7F	260	1995
Tampa Electric IGCC/Florida, USA	Texaco gasifier	GE 7F	250	1996
Elcogas IGCC/Puertollano, Spain	Uhde PRENFLO	Siemens V94.3	330	1998
Nakoso IGCC demo/Fukushima, Japan*	MHI gasifier	MHI 7 frame	250 (gross)	2007
Huaneng Greengen IGCC/Tianjin, China	Huaneng gasifier	Siemens V94	250	2012
Duke Energy IGCC/Edwardsport, Indiana, USA	GE gasifier	GE7FB	618	2013
Kowepo IGCC/Korea**	Shell gasifier	GE 7F?	250	2016
Kemper IGCC/Mississippi, USA***	KBR TRIG	Siemens V94	524	2017
Nakoso IGCC/Fukushima, Japan	MHI gasifier	MHI 7 frame	540 (gross)	2021

Note *The Mitsubishi gasifier is a two stage air blown, instead of oxygen, gasification designed for power generation. ** source of information (AO Technology, 2012) (power Magazine 2021). *** Southern company announced to abandon the ongoing commissioning of gasification island, which is an air blown gasification, in 2017

appears that this trend would continue moving forward, primarily driven by the demand for a high gasification efficiency that a dry feed modern gasification could offer, which is encouraged by the central and local governments. And last, gasification projects have become a kind of standardized on a large scale and formed their own eco-environment in an industrial park setup based on the extensive hands-on experiences amassed up in previous decades. For example, since the first coal to olefins project was put into service in Inner Mongolia in 2010, producing an annual 600,000 tons of olefins from bituminous coal, there have been 17 projects of the same scale or larger currently in operation or under an advanced development. These projects have a combined capacity of up to annual 8.7 mm tons of olefins, a supplementary source to the same products of a crude origin. These projects use similar bituminous or subbituminous coal as feedstock. After demonstrating three Fischer–Tropsch projects at the level of up to 160,000 tons of liquid fuels around 2008–2010, there have been 7 more such projects being executed among which 6 of them are designed to produce annual 1 million or more tons of liquid fuels from bituminous or subbituminous coals. The largest with an annual capacity of 4 million tons of liquid fuels has been in commercial operation in Ningxia since 2016. The pressure entrained slagging gasification process has demonstrated its value, the quality of syngas, and the required availability and reliability. With the same token, the coal derived chemicals or liquid fuels have proven as viable supplementary sources of products that have typically been produced from crude oil. It has been claimed that the liquid fuels would be competitive as long as the crude price stays around $50–60/bbl.

From the perspective of the synthetic industry, the modern entrained gasification operated under a slagging and pressure conditions proves it the right technology, replacing the water gas process, and is capable of delivering the required quality and quantity of syngas that is competitive enough to make the synthetic chemicals viable under certain market conditions. Such an achievement, however, has taken a long journey and would not be possible without several critical developments. On a macro level, the modern coal gasification technology can in general be characterized by the measures such as a continuous operation, a high intensity, a large scale, and a pulverized coal fed system. First of all, the separation of oxygen availed by ASU technology certainly makes possible the leap of gasification technology from a non-continuous operation as with the water gas process to a continuous one. Then, process-wise, the combined outcome of a slagging temperature, a pressurized operation, the injection of small particle fines of coal (wet and dry), and again the use of oxygen has enabled the modern gasifier with a high intensity and high capacity, another leap from previous gasification processes of both water gas generators and gas producers. The underlined chemistry is obvious that a slagging temperature and the uses of small particle fines of coal and oxygen would expedite the gasification reactions so to break down any large hydrocarbon molecules into small ones in a matter of a single digit of second, and the high pressure would take the advantage of the compressibility of gases to make the gasification process compact.

Looking back at the historical development, the innovation of gasification technologies has in many occasions been driven by feedstock, which is reflected by the fact that the experience of the past 230 years has proven repeatedly that the properties of coals, physical and chemical, have a significant impact on gasification technology, engineering, and operation. A successful project always hinges on a good understanding of the specific coal that is to be chosen as feedstock. This is because the coal and any coal are highly heterogeneous and their properties vary so broadly depending not only on its origin and location but also on its geological conditions over the millions of years during which foreign materials such as rocks, dirt and minerals could become part of it as ash. In addition, the mining method also has additional impact on the ash as well. Understanding the ash behavior such as melting point and viscosity etc. has, therefore, become necessary for the operation of any of the modern gasification technologies. Engineering wise, it is Dr. Schmalfeldt's pioneering work to gasify coal fines in an entrained fashion that kick started the course of the modern gasification technology, and then the additional works laid by the slagging gasifiers, the Winkler gasifier, and the Koppers–Totzek gasifier had no doubt demonstrated from different perspectives the potential values of a modern gasification technology. To break the last piece of ice, it is the Texaco's slurry gasification that scored an important milestone to make the pressure work for the gasifier, instead of against it. Then, it takes Texaco about another twenty years to firm up the necessary techniques before putting it into demonstrations during the 1970s and 1980s.

Through commercial projects in the next half century, the modern gasification processes continue to evolve by expanding the feedstock portfolio of coals and petroleum coke. Generally speaking, nowadays, a wider range of coals from subbituminous to anthracite, regardless of their caking nature, and petroleum coke should have no technical difficulty to be gasified by an appropriate pressure entrained gasification technology, no matter wet or dry feeding system, for syngas production. This is, however, by no mean to say that the modern gasification technology would be able to handle all types of coals. For coals such as low rank coals containing high moisture content and some special coals exhibiting an unusually high ash melting temperatures such as the Changzhi coal, however, exercising an additional precaution would help minimize the risks associated with the selections of gasification technology and the related process etc. The lessons learned so far tell a simple fact that these types of feedstock tend to perform poorly in a modern gasifier due to the additional energy to deal with either the high moisture content or the required operating temperature higher than typical, which the latter case would also result in higher wear and tear to the internal parts of a gasifier. For cases that have to be, the Lurgi gasifier may well be the appropriate choice to go. In such a scenario, developing potential synergies for the by-products generated would typically help the overall project economics in the right market. Last, for coals containing high impurities including chloride and alkaline metals like sodium and potassium etc., regardless their ranks, it should be mindful in engineering and operation; or if possible, these coals should be avoided should there be no prices of products to support the increased capex and opex.

The early industrial deployment during the 1980–2000 worldwide acquired necessary information on engineering the modern gasification and certain critical equipment. Then, the subsequent massive deployment since 2000 has provided the wanted opportunity for gasification to grow full-fledged and become mature. In so doing, there also established a valuable ecosystem for the gasification business as a whole (Zhuang, 2015). Such an ecosystem includes the necessary infrastructure ranging from technology, engineering and construction, equipment manufacturing, and related logistics to the wealth of hands-on experience and expertise from operating a wide range of feedstock portfolio, and services around the commissioning, operation, and maintenance of a gasification facility, all proving priceless for the development of ongoing gasification projects. Such an ecosystem, for example, has fostered the turnkey business model back on track, and many engineering companies are no longer hesitant to take an EPC contract. Moreover, some companies and developers are willing to take the build to own business model around the gasification island by selling syngas over the fence. Coal gasification technology appears to have grown mature but, nevertheless, not without limit.

Believe it or not, gasification is still in a state of the art mode and not able to be predicted mathematically. The wealth of hands-on experiences, knowledge, and know-hows on which gasification has been built are, therefore, still reminiscent, relevant and applicable in principle to many cases and problems faced by projects nowadays. What Clegg Jr. stated about the necessary practice and required skillset for coal gas making still echoes (Samuel Clegg, 1841).

The retorts, for instance, must be set in such a manner as to be heated sufficiently with the smallest quantity of fuel, and without the liability of being made too hot; they must be carefully watched, lest the incrustation of carbon in the interior accumulate more than is absolutely unavoidable, and lest they burn out before the time calculated. These particulars can only be learned by practice; they will vary with every different quality of coal, and be affected by the size and shape of the distillatory vessel itself. It also requires much experience to determine the most economical arrangements for the condensers, purifiers, etc., and to regulate the quantity of gas to the demand.

References

Alpert, S. (1986). IGCC phased construction for flexible growth. *EPRI Journal*, 4–12.

Anastai, J. (1980). *SASOL: South Africa's oil from coal story*. US EPA.

AO Technology. (1981). *Japan's sunshine project solar energy R&D program*. Ministry of International Trade and Industry.

Burgt, C. H. (2003). *Gasification*. Gulf Professional Publishing.

Burgt, E. A. (1981). Development of the Shell–Koppers coal gasification process. *Philosophical Transactions of the Royal Society of London*, 111–120.

Clusen, R. (1980). *Final environmental impact statement—great plains gasification project*. USDOE.

Corp, T. B. (1980). *Coal gasification systems engineering and analysis*. NASA.

Corp, R. (1983). *Environmental characterization of the Texaco coal gasification at Oberhausen*. EPRI.

Dierschke, A. (1955). Development of the coke oven industry. *The Fuel Society of Japan*. The Fuel Society of Japan.

DOE, U. (2006). *Practical experience gained during the 1st twenty years of operation of the GPGP*. USDOE

Elliott, C. G. (1963). *Chemistry of coal utilization*. Wiley.

Erasmus, M. D. (1987), (12). Update of the sasol synfuels progress. *The American Revolution: Energy* 1–21.

Hongjian Jing, G. W. (2000). *Practice and study of the Lurgi pressurized gasification for ammonia production*. China Chemical Industry Press Co., Ltd.

Moor, F. (1953). Patent No. 1655443. USA.

Partridge, G. C. (1982). Coal slurry pipelines: The ESTI project. *Right of Way*, 11–16.

Samuel Clegg, J. (1841). *A practical treatise on manufacture and distribution of coal-gas*.

Slater, W. S. (1970). Patent No. 3544291. USA.

Stroud, D. H. (1981). *Chemistry of coal utilization*. Wiley.

Technology, P. (2012). *Kowepo selects GE technology for IGCC power plant*. Retrieved Feb 2023, from Power Technology, http://power-technology.com/news/newskowepo-selects-ge-technology-for-igcc-power-plant/

Whipple, D. (2014). *Coal Slurry: an idea that came and went*. Retrieved Feb 2023, from WhyHistory.Org: https://www.wyohistory.org/encyclopedia/coal-slurry-idea-came-and-went

Zhuang, Q. (2015). An overview of gasification commercialization for power. In *Gasification technologies conference 2015*. Gasification Technologies Council.

The Journey of Energy Transition

14

The evolution of coal gasification technology presented a closer look into the industrial revolution across the nineteenth century and the one that followed. It is a journey not in any shortage of the much needed inspirations to technology innovation and the search for a viable solution toward the energy future of our world. In retrospect, coal gas illumination that lighted up the early part of the industrial revolution, or the so called first industrial revolution kicked off by the deployment of steam engines, is an important milestone in human civilization. Then, the subsequent diversification of the coal gasification process had enabled multiple pivotal technologies that triggered the second industrial revolution typically characterized with steel and gas engines, which had created profound impacts to almost every corner of human society. Such impacts are still going on today. On the flip side of the positive social and economic development, however, the accumulative changes and impacts that human activities had made so far to Mother Nature has added up to today's debate about climate change and the subsequent damages and loses to economies, human lives, and the biodiversity of the global ecosystem. In search for viable solutions for the needed energy transition to fight against the increasingly extreme and unpredictable weather as a result, it appears that hydrogen resurfaces as an attractive source of energy that could be part of the viable solution to counter the climate change. This would present the coal gasification a great opportunity to exert its impacts and functions because the gasification process is not only one of the largest sources of hydrogen, if not the largest one, but an effective process to decarbonize fossil fuels, which would provide our society ample time as an insurance policy, the least, to make the necessary transition into the future world of a zero or net-zero emission of carbon. Achieving a net-zero carbon emission economy simply needs time, a time with no back end yet in sight.

© The Author(s), under exclusive license to Springer Nature Switzerland AG 2024   225
Q. Zhuang, *From Coal to Hydrogen*, Synthesis Lectures on Chemical Engineering and
Biochemical Engineering, https://doi.org/10.1007/978-3-031-55586-2_14

Energy transition is nothing new at all. It actually started from the time when coal gasification technology was born by turning coal into clean fuel gases full of chemical energy. In essence, the evolution of the coal gasification technology itself mirrors a long journey of the energy transition when there were no other alternatives since the onset of the industrial revolution. To grasp such transformation that coal gasification had gone through over the past 230 years and how it played into the Industrial Revolution and the modern industry, coal gasification can be reexamined from a few different perspectives.

The Road to Modern Gasification

From the viewpoint of evolution itself, it all started in 1792 when Murdock used the coal gas that he cooked out in his backyard at Redruth to illuminate his home, which immediately became the most popular tool during the early part of the Industrial Revolution by extending daytime into dark evening. What had brought the ultimate light to the society as a whole is the public utility model that was pioneered by Winsor who believed that the public places was where the new lighting would be needed the most. It turned out to be a great entry point for coal gas into the market, which was followed by a rapid propagation. Overcoming the initial hurdles of all sorts and aided by setting up a few gas lights on Paul Mall Street in 1807 to showcase the new gas lighting, Winsor was finally granted the needed license in 1810, and on the New Year's Eve of three years later, Londoners first time witnessed a brighter Westminster Bridge. Soon, coal gas makes its way into many other cities, towns and villages of England, and enjoyed an unmatched growth in the following century. Coal gas reached anyplaces wherever an economic development would take place. The second major milestone for coal gasification arrived when the growing locomotive business demanded steel so to make its rail tracks last longer around the mid-nineteenth century. Coal fired metallurgical furnaces typically used at the time could not make the needed quality and quantity of the steel. It is the open hearth furnace that Siemens brothers invented during the 1860s that is capable of manufacturing a large quantity of good quality steel. Compared with the Bessemer process already industrially available at the time, the open hearth process is much more robust, flexible, is able to process both pig irons and scrap metals directly, and therefore, delivered the steel products that the market demanded at much lower prices. How happened? Well, the Siemens brothers made the miracle by dusting off an old gas producer technology and put it to work with the regeneration furnace that they invented back in the 1850s. The integration between the gas producer and the regeneration furnace had removed the bottleneck of reaching a high temperature faced typical coal fired furnaces, and opened up the door toward the second industrial revolution.

The coal gas business in England during the 1860s continued its course of fast growth while penetrating as a clean fuel gas into the households for cooking and heating. Competitions from inside and outside the coal gas business also became intensified as well, which forced gasworks to look for ways to make their operations more cost effective. Becoming the victim of its own success during a time of change, the traditional coal gas business after a long prosperity unfortunately entered into a mode of consolidation or reorganization as a way to make its operation more cost effective. Although the producer gas expanded as a fuel gas further into industries of metallurgy, glass making, calcining, and so forth, the low BTU producer gas has no values to the traditional coal gas business. England started to feel the pain inflicted by its stagnant coal gas making technology. During the 1870s, a brand new coal gasification that extended the age of coal gas illuminating brought the tentacles of technology innovation to the North America. It was the water gas process invented by the civil war veteran and aeronaut, Thaddeus Lowe. Lowe was not strange to coal gas and its making at all. He used coal gas-filled balloons multiple times to send himself up in the air for adventure flights prior to the war, and to conduct aerial reconnaissance for field intelligence on Confederation Army during the war. After the war, Lowe threw himself into the technology business by applying new scientific discoveries at that time. The water gas process as one of Lowe's inventions had become one of the most important developments in coal gasification technology. The water gas proves as a flexible coal gas, not only extending the age of the coal gas illumination but also becoming the only viable source of hydrogen when the Haber–Bosch process delivered the first artificial ammonia decades later.

The water gas generator is in principle not much different from the coal gas producer utilized by the Siemens. With a several smart twists made to the operating procedure, Lowe was able to produce a completely different, high calorific value, and high-quality synthesis gas or syngas. To make an illuminating gas, Lowe injected the cheaply available naphtha or light oil in the North America market into the hot water gas stream, which becomes a carburetted water gas, a quality illuminating gas containing enough olefin compounds and having a heating value of 550 BTU/cubic feet or higher. A few decades later, in addition, the water gas would find its great value when synthetic industry emerges. In the meanwhile, another complete gasification technology of coal was born.

In England from about 1880, the traditional coal gas making can be characterized by the extension of new technologies and their integration. Coal gasworks started to improve its efficiency and operation by applying the integrated gas producer that the Siemens brothers invented. Before 1889, upgrading coal gasworks with the water gas process had remained to be a phenomenon of the North American market. Then, the water gas process (already owned by UGI) was introduced into England to help the Beckton gasworks meet the gas demand during a peak time, and many more followed the suit later. Overall, the traditional coal gas making remained essentially unchanged until the vertical retort process emerged during the early 1900s. Around the time, Mond who entered the coal gasification field inadvertently by trying to fix Air-N to make ammonia ventured down the path and worked out the largest gasifier that successfully processed a cheap local caking bituminous coal. The cost effective Mond fuel gas had made the Mond gas an affordable competitor in the large engine market including that for the generation of electricity. By then, the producer gas had deeply penetrated every corner of industries, which is another significant step forward in the energy transition by eliminating those polluting chimneys of the many coal fired boilers. Producer gas in general continued its services until post WWII in the US and the 1960s in England and most part of Europe.

The groundbreaking at Appau with the synthesis of ammonia in 1913 opened up a completely new market for water gas. Then, the subsequent synthetic methanol in 1923 and the synthetic plants for liquid fuels via either hydrogenation process or Fischer–Tropsch process around the mid-1930s had made the water gas indispensable in the production of the high quality syngas required by the synthetic processes. Such a syngas, rich in hydrogen and carbon monoxide and minimum in inert gas, would be fed into the synthetic processes after being cleaned up to remove acid gases and other impurities. To meet additional demand for syngas as part of the war effort in Germany, a large amount of resources were poured in to the innovation of gasification processes for the purpose to expand its feedstock portfolio to less desired feedstocks such as coke fines and low rank coals. To accommodate the low quality coals, extensive innovations took place at several plants to demonstrate the working gasification technologies between the mid-1920s and the breakout of WWII. The old slagging gasification that was invented by Ebelmen was pulled out to serve the new needs to digest process refuse. Syngas making had become creative as well such as COG was cracked separately or in a water gas generator environment to supplement additional demand for hydrogen. At several other plants, more comprehensive technical schemes were contrived to process the low rank coals. Each of the comprehensive processes features a vertical retort that was integrated with a water gas generator where the most needed heat was provided with a regenerative furnace. The Koppers-Spuelgas water gas generator, the Pintsch-Hillebrand water gas generator, and the Didier-Bubiag water gas process are examples of the endeavor. Gasification technologies of brand new principles were invented as well such as the fluidized bed Winkler gasifier that was designed to handle coal fines or coke breeze, and the entrained Schmalfeldt integrated water gas generator, which took in as received low rank coal as feedstock and gasified it to completion. Each of these versatile processes has its own merits and drawbacks but it appears that all were managed, of course with different levels of success, to produce the needed syngas for the production of liquid fuels under the circumstances at the time. By looking carefully into these comprehensive special processes, it is noticeable that what makes these special processes complicated as they needed to be is the ways to transfer as effectively the required intensive heat as possible to sustain the endothermic water gas reactions (Chap. 10, **Reaction 4**, **5** & **6**) that would produce the high quality syngas for synthetic processes. Speaking of the effective heat creation inside a gasifier, the air separation technology that separates oxygen from air rendered coal gasification another significant step forward to become a continuous process for the production of syngas. Examples are the Winkler gasifier, the Wuerth slagging gasifier, and the Lurgi gasifier that were designed to use oxygen as the oxidant to produce the syngas continuously. The available oxygen supply becomes the key to essentially combining a gas producer and a water gas process into one. At the end of the war, the information about these technologies, processes and plant operations had been collected by the Allied forces and become a valuable asset down the road.

The shift of energy market post WWII pretty much left the costive synthetic liquid fuel facilities in Germany behind. Except for the ongoing demand for the production of ammonia worldwide, activity around coal gasification remained low. Gasification processes such as the water gas generators, the Winkler gasifiers, and the Lurgi gasifiers were built sporadically here and there globally. In the meantime, governments like the US and some private companies, however, took a strategic approach to coal gasification and synthetic liquid fuels. Surrounding what gasification technology would better fit the future demand, development efforts appear to be laser-focused on the principles of the entrained gasification that Dr. Schmalfeldt first developed before WWII. Design variables among several gasifier designs under development at the time were narrowed down to a few critical parameters such as ways to feed coal fines into the gasifier, operating pressure, and temperature etc. The Koppers-Totzek gasifier is one process that was developed under the entrained principle and deployed for ammonia production in the early 1950s. It was a bulky system of an atmospheric slagging operation but, nonetheless, served the global ammonia market well at the time. To make it compact, however, a gasification process needs to go pressurized. Here, a solution became necessary to effectively feed coal fines into a pressurized gasifier. The Texaco gasifier was developed to do just that; it mixes coal fines with water into slurry, which is pumped conveniently into the Texaco gasifier. Then, Texaco Inc. placed the technology into an industrial demonstration in the 1970s and successfully deployed it in the decades that followed. The era of modern coal gasification began to evolve slowly till this day. Looking back, what modern coal gasification has achieved to turn a solid fuel into coal gas for different purposes either as illuminating gas, fuel gas or syngas is obviously significant over such a long journey during and after the industrial revolution, a journey that is practically impossible to accomplish without the passion and ingenuity of the many great minds in addition to the chemistry that has been established along the way.

The Perspective of Chemistry

Scientifically speaking, coal gasification as one of the few earliest chemical process technologies would hardly be separated from the development of science but rather goes hand in hand with chemistry from its onset. First of all, the recognition of the gas state and the establishment of the combustion theory by Lavoisier provided the needed clarification of the confusing 'air's and as well the demystification of fire or combustion phenomenon, which at least helped ease the unnecessary fears when Murdock and Winsor brought the coal gas lighting into textile factories and the public space, respectively. Although the coal gas business enjoyed astronomical growth in the next half a century the coal gas making technology had, however, somewhat become stagnant. So did the chemistry. From the perspective of coal gasification, chemistry during the most part of the nineteenth century lagged behind several other scientific fields. Coal gas making certainly felt the pains and struggles to find necessary supporting fundamentals to look at for the purposes either to improve itself or to understand and interpret what is going on inside the retort. Engineers and inventors, nonetheless, plowed ahead anyway. Such a moment finally arrived when Joule demonstrated the different forms of energy and their interchangeability during the 1840s, which removed the last veil about the caloric that Lavoisier left in his chemistry theory and eventually led to the establishment of the first law of thermodynamics during the 1850s.

Joule's work also cleared the clouds hanging over for so long about concepts of energy, heat, mechanical movement, electricity, magnetic force and so forth; they are just different forms of energy and interchangeable. Such a breakthrough in chemistry inspired many who had been working to improve the efficiency of steam engines and fireplaces in order to save coal consumption. Among them are the Siemens brothers who worked in different fields but shared the same focus of how to better use heat. When the young Frederick invented regenerators to recover heat from hot exhaust flue for glass making, the elder William immediately realized the value of how heat could be effectively transferred. The brothers had ever since worked together and created the open hearth process in the 1860s that revolutionized steel making industry, which was pivotal to the industrial revolution at the time. What allowed them to do so is the old gas producer that was invented by Bischof two decades back to make a producer gas. Differentiating from the traditional coal gas making by retorting, the gas producer under the new chemistry is a continuously operated gasifier that uses air blast through a hot coal bed so that coal reacts with oxygen in the air to its completion. What comes out of the gas producer is a fuel gas of low heating value due to the large nitrogen presence in the air. The low heating value fuel gas, once being heated in the regenerative furnace, becomes a powerful source of heat for steel making. Along with many other developments and discoveries in both the scientific and industrial worlds, the new chemistry taking place in a gas producer called for a fundamental clarification because the popular Dalton atomic theory faced more dilemmas than difficulties. In response to such changes and the renewed interest in coal tars to make dyestuff, the debate over molecular theory proposed by Avogadro in 1811 resurfaced around the 1860s, if not earlier, and started to take root. When van't Hoff proposed the valence bonding theory during the 1870s that supported the molecular theory, it also provided the needed ammunition at the time to satisfy engineers with a simple and consistent explanation that allowed the dissection of the gasification reactions taking place either in a gas producer or a water gas generator and find ways to improve it. Dowson's work during the early 1900s certainly benefitted from such a development in chemistry. His depiction of what happens inside a producer and the modeling under certain assumptions did provide a clear picture the first time on a micro level and has been followed by many future R&D activities. To allow Dowson to carry out the modeling, however, another important milestone in chemistry can't be ignored. It is the second law of thermodynamics jointly developed by Joule and Sir Kelvin during the 1850s. Their discovery that the temperature of a gas would decrease when being allowed to expand without doing work led to the establishment of cryogenic technology which is critical to coal gasification down the road in many ways. First, physicists and chemists or physical chemists were finally able to isolate many more individual gases that were impossible before. They were now able to study the thermodynamic properties and the reaction characteristics of each of the individual gases. Such information is important for a thermodynamic analysis of a gas producer as Dowson did. Secondly, the cryogenic principle also led to the development of the ASU technology around 1900 that separates oxygen from the air, which becomes the another step to revolutionize coal gasification several decades later and has ever since become widely practiced in the industry of the modern day. With the evolution of coal gasification going on, chemistry has become an inseparable part inside out.

De-Carbonization

From the viewpoint of the industrial revolution, the importance of coal gasification has been many folds, deep, thorough and far-reaching. First, the coal gas illuminated almost every perspective of human society, dramatically improved the public lives and safety, and boosted the productivity of the economy as a whole by extending daytime into night. Such importance continued for about a century until electricity came to market around WWI. Then, the producer gas enabled open hearth process unleashed the production capacity of the mostly needed steel material for rail tracks, boats, sea-going ships, and structure steel for high skyscrapers etc while reduced the cost of it. Around the same time, the Belgium inventor Lenoir fired up coal gas in a converted double acting steam engine; here he created a two stroke gas engine running smoothly. But Lenoir did not use the engine to make a standalone automobile, a dream that was tried almost a century ago in France, even though he tried. He sold hundreds of his two stroke engines, instead, to small shops and works as stationery mechanical drivers. This is because coal gas was the only practical, available, and easily accessible fuel in cities like Paris and many others where coal gas was distributed. Once Otto improved Lenoir's design into a four stroke engine in the mid-1870s, gas engines had become much more powerful and more efficient, which attracted many more small shops and factories. Otto and his partner sold thousands of them, again, as stationary prime drivers. Such a wide and rapid deployment of gas engines provided the badly wanted mechanical drivers to almost every corner of industrial and commercial fields that steam engines could not reach or compete with. More has yet to come.

Soon, two of Otto's engineers, Daimler and Maybach left the company to pursue the century old dream of building the *fardier*. They succeeded in 1890 by tapping the then available liquid fuel, gasoline, which quickly revolutionized the transportation world. Back to the gas engine business, there was a new development that made the operation of gas engines possible off the coal gas grid, extending the gas engine market to wherever the needs were. The gas engine business had ever since entered into a fast lane of growth and become the indispensable prime driver during the ongoing industrial revolution. It is aided by the compact gas producer that Dowson developed, and was further amplified when Mond designed a much larger gasifier to deliver a large quantity of the coal gas at much more competitive prices. The domino effect of coal gasification continues to roll on and on like an ice ball until hitting another jack box in 1913 when synthetic chemistry successfully produced the first ammonia by fixing Air-N with hydrogen, followed by the synthesis of methanol, and then liquid fuels. A new industry was born, the synthetic industry. Water gas had extended its role as fuel gas and illumination gas to raw materials for the new industry. Such a shifting provided an opportunity for coal gasification or gasification process in general to implement itself with relevant downstream processes for the production of hydrogen or syngas during which carbon dioxide is typically conveniently processed as a concentrated stream, a necessary step in the process.

Speaking of the energy transition, it essentially began when coal gas making started to turn coal into coal gas, first as a far superior illuminating gas to light up streets, houses, and buildings etc. and then as a clean fuel gas to fire up boilers, furnaces, and gas engines that had impacted the industrial activities so profoundly and long lasting. Since the inception of steam engines from the early eighteenth century coal has become deeply rooted in the industrial development in the Western world till this day, and coal gasification has been working along side with the coal combustion to meet various industrial demands. According to BP's energy statistics, fossil energy combined (oil, natural gas and coal) in 2022 made up 81% of the total world consumption of primary energy, remaining unchanged from the previous year but reduced by 4% from 2019 before the Covid-19 hit the world. Specific to coal, it made up 27% in 2022, which remained unchanged from 2021 but reduced by 1% from two years earlier. The balance is nuclear energy, hydropower and renewable energy among which the renewable energy reached 7% of the total energy used for 2022.

Comparing to 1900 when coal was the dominant industrial fuel source (Fig. 14.1), the current world primary energy structure has absolutely become more diversified and more risk tolerant by reducing its reliance on fossil energy especially coal. Such a diversification or transition over the past 120 years has taken many steps in the making by looking closely into it. Generally speaking, the dominant position of coal as the primary energy did not change until the end of WWII. Wars had significant impact to the demand for coal, up and down dramatically. As the largest industrial power of the world since 1900, for example, the US produced near 600 million tons of coal annually during WWI, then dropped almost to 50% at the time of the great depression. The production picked up again during the 1930s and peaked over 600 million tons of coal annually at the end of WWII. The demand for coal post the war slipped back and decreased to below annual 400 million tons during the 1950s (*based on data from the USGS website*). Coincidentally, commercial activities on coal gasification also reached its low time as well. Relative to

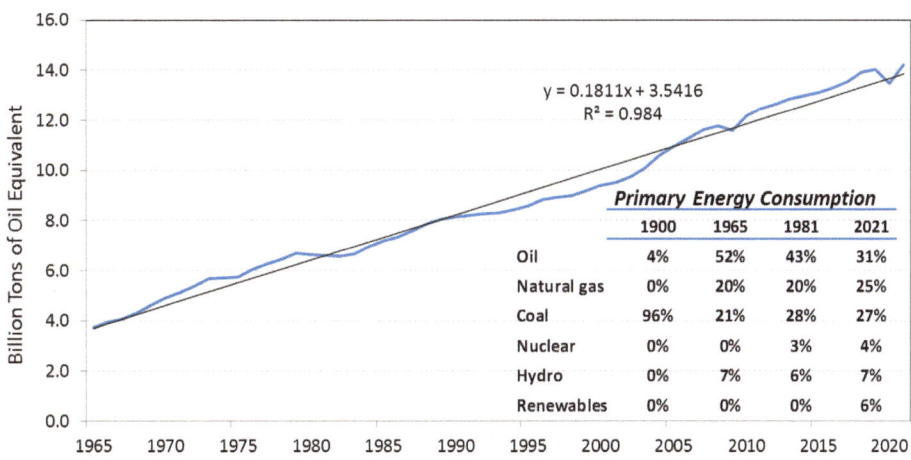

Fig. 14.1 Post war worldwide primary energy consumption (Per BP Energy Statistics; coal data for 1965 was an estimate and that of 1900 from Fischer report obviously excluded biomass use)

coal, crude oil production in the US during this period had steadily increased and eventually overtaken coal as the primary energy source on the account of both mass and energy equivalence. From the early 1960s, the post war recovery started to pull more coal back to the market. Coal production in the US reached 470 million tons in 1965, roared back up to about 900 mm tons in 1990, and hovered at the level before beginning a gradual downward trend till around 2015.

The total primary energy consumption worldwide increased steadily along a straight line to about 350% between 1965 and 2020 (Fig. 14.1). Although BP started to publish its energy statistics from 1965 coal data was left out until 1981. Between 1981 and 2021, it is obvious that coal, nuclear, and hydro energy remained little changed; the decrease on oil consumption was mostly made up by the increases of natural gas and renewables, which the latter primarily stemmed from the development of solar and wind energy in the past decade. To make use of the BP's information for the purpose of grasping a whole picture of 1965, let's assume that the US consumed the same amount of coal as the rest of the world for the year. On oil equivalent basis, fossil energy as a whole was dominant at 93% in the primary energy structure worldwide where crude oil contributed more than 50%. This estimate or guess looks certainly low for coal in the primary energy at the time according to the World Energy Consumption of 1800-2000 by the Encyclopédie de l'énergie, which coal use stood somewhere between 38-27% during 1960 and 1970. The number would further be diluted by the consumption of biomass for the year, which should be much higher during 1900. Adjusted by the large biomass consumption back then, coal in 1900 would probably be around 45-48% in the overall primary energy structure worldwide. The downward trend for coal use from 1900 remains unchanged, that is, coal share in the average global energy portfolio has been decreasing till this day.

Although started in the 1950s, the commercial development of nuclear energy for electricity generation did not take place until 1965 and a substantial number of nuclear power plants were actually commissioned during the 1970s. Spurred by the oil crisis from 1973, the 1980s experienced an expedited growth of nuclear capacity for the generation of electricity. The development in the oil and gas field between 1900 and 1965, first in North America and then followed by the post war discoveries in the Middle East and the North Sea, made the energy transition possible with the alternative fuels of oil and gas. Like the coal gas network developed during the nineteenth century, the post war pipeline network for natural gas distribution certainly played into not only phasing out coal gas but also increasing its share in the overall primary energy portfolio. Concerning the energy transition, the hydro power capacity that has been in service from the 1910s should not be ignored at all, but rather one of the preferred renewable energy sources.

Considering the magnitude of primary energy consumed overall each year, the transition that has taken place within the primary energy landscape over the past 230 years was not small by any measure. There appears a several ways to view such an energy transition depending on what kind of lens to look through. In general, the landscape of the primary energy transitions toward low carbon intensity by significantly reducing the share of coal from its absolute dominance to 27% in 2021, a drastic shift, and the rest is fulfilled with 56% of crude oil and natural gas and 17% of nuclear, hydro and renewables. Although coal gave up its dominance in fossil energy fossil energy still collectively contributes a lion's share of 83%. An inconvenient reality is that our world economy is dominated by fossil energy as a whole and the ongoing global economic activity is still reliant on fossil energy including coal. The journey to achieve a net zero of carbon emissions would hardly be a short one without effective wherewithals.

BP in its outlook 2020 projected a scenario of net zero of carbon emission to be secured by 2050, which would require 59% of renewable energy, 10% hydro and 9% nuclear energy, a total of 78% in the energy mix. Simply thinking of how long it has taken to get us to where we stand today, 30 years of time would seem not long and it would require at least an exponential growth of renewable, hydro and nuclear energy from current 17% to 78%, a 61% increase. In addition, coal would need to go down to 2%. It is an ambitious goal.

The breakthroughs made in technologies cross energy field in the recent decades certainly look encouraging. Examples are that products of solar panel and wind turbine have become much more efficient and competitive. The advanced nuclear technology such as AP1000 and its commercial deployment in China in the past decade and the current commissioning of two AP1000 reactors at Vogtle, Georgia in the US are another promising step in the right direction. Saying it promising, however, does not necessarily mean without problems and challenges. The disasters took place at the Three Mile Island of the US in 1979, at the Chernobyl of the former USSR in 1986, and at the Fukushima Daiichi, Japan in 2011 are the constant reminder of how important the safety should be an integral part of the nuclear solution. In addition, the management of nuclear wastes should be adequately addressed as well, foreseeing the potential increase in the deployment of more nuclear energy. The setbacks in the development of the nuclear waste processing facilities both in Japan and the US are, however, not promising at all, but rather like additional salts poured to the wounds of the industry. The Rokkasho nuclear waste reprocessing plant located in northeastern Aomori Prefecture, Japan is designed to process 800 tons of nuclear wastes annually. Although its construction was started as early as in 1993 the future of the plant is still obscure, according to the World Nuclear News in January, 2023, at best. Back to the US, what happened to the Yucca Mountain Nuclear Waste Repository project, received the regulatory approval in 2002 but then dropped in no more than 10 years later, is another example, which certainly cast a significant amount of shadow about the future potential that nuclear technology could offer. When it comes to solar and wind technologies, a reliable back up plan, when there is no sunshine or wind, would have to be part of the solution in addition to maintaining the needed stability of the electric grid. While battery technology currently under development will help smooth out the flow of electrons other technologies such as hydro power and thermal power would still need to be handy when necessary. This may require some fundamental shifts when utility sector structures its contingence plan or plan of reserve capacity to fill the anticipated shortage in supply, which could be created by the shortage or an extended one by Mother Nature. The thermal power plants are typically powered by coal, oil or natural gas etc. Facing current status quo of 83% fossil energy including 27% coal in the primary energy landscape, transitioning into a hydrogen fueled economy would hardly be possible without the right technologies. Here, gasification technology including partial oxidation, steam methane reformer and the like when dealing with liquid and gaseous hydrocarbons as feedstock will come handy.

In fact, gasification technology is not only capable of converting coal into a clean fuel gas and a syngas but also an effective tool to manage carbon emissions in the utilizations of coal and other hydrocarbons. How? When coal gas evolves from a fuel gas to a syngas for chemical synthesis de-carbonization had already taken place, both inside and outside a gasifier. Depending on the final product that would be produced from syngas, carbon in coal in general would end up in two major streams, a portion being fixed into the product and the rest being rejected as a stream of concentrated carbon dioxide. Taking the recent industrial coal to methanol application as an example, about a third of the carbon in coal that is fed into a modern gasifier is fixed into methanol product and about 60% of the carbon could be rejected as carbon dioxide of up to 90% concentration depending on the applied process for gas processing and its operating conditions. The balance of the carbon went to coal ash as unconverted carbon and other small streams around the plant. From the standpoint of coal gasification, or process engineering in general, capturing carbon dioxide is not an issue at all but rather a done deal already. The issue is where to find a permanent home for the concentrated stream of carbon dioxide and thereof the associated cost to carry it out. To this development, many may have wondered what a waste it would be to capture carbon dioxide of about 400 ppm or 0.04% from the thin atmosphere on the one hand while simply letting go of the concentrated carbon dioxide at the stack back to the atmosphere in many gasification plants on the other. Unlike solid waste handling, carbon dioxide is a global phenomenon; once reaching the atmosphere, the molecules of carbon dioxide will soon be evenly distributed to every corner of the earth. Like winning a war, being preemptive and capturing carbon dioxide at the sources should always be the first position to take as a strategy. Of course, there are such kinds of positive development in recent decades such as carbon sequestration by injecting carbon dioxide underground and developing catalysts that convert carbon dioxide back to the value-added products. It is a good start but far from being enough if the world is determined to become a net zero on carbon emissions.

References

Barraclough, K. C. (1981). *The Development of the Early*. University of Sheffield.

Binder, F. M. (n.d.). *Pennsylvania Coal: An Historical Study of Its Utilization to 1860*. Philadelphia: University of Pennsylvania.

Clark, D. K. (1880). *Fuels, its Combustion and Economy*. Crosby Lockwood and Co.

Co., D. G. (n.d.). *About Us*. Retrieved Feb 2023, from Dakota Gasification Company: https://www.dakotagas.com/about-us/index

Coal Gasification. (n.d.). Retrieved from https://en.wikipedia.org/wiki/Coal_gasification

French Aerostatic Corps. (n.d.). Retrieved from Wikipedia: https://en.m.wikipedia.org/wiki/French_Aerostatic_Corps

Frederick, W. (1862). On a regenerative gas furnace. *Proceedings of the Institute of Mechanical Engineers*, (p. 21).

German Culture. (n.d.). Retrieved Nov. 2022, from Carl von Linde Who Gave the World the Refrigerator: https://germanculture.com.ua/famous-germans/carl-von-linde-who-gave-the-world-the-refrigerator/

Gas Making Technology. (n.d.). Retrieved mar 1, 2022, from Chronicles of Shanghai Sciences & Technology: http://61.129.65.112/dfz_web/DFZ/DulanMu?idnode=4454&tableName=userobject1a&id=-1

Grace's Guide Ltd. (2021). *William Siemens*. Retrieved May 30, 2023, from Grace's Guide To British Industrial History: https://www.gracesguide.co.uk/William_Siemens

Grace's Guide to English Industrial History. (n.d.). Retrieved from Grace's Guide: https://www.gracesguide.co.uk/Humphreys_and_Glasgow

Guide, G. (n.d.). *Gas Light and Coke Co*. Retrieved March 2022, from Grace's Guide to British Industrial History: https://www.gracesguide.co.uk/Gas_Light_and_Coke_Co

Guide, G. (n.d.). *Joseph E. Dowson*. Retrieved from GraceGuide: https://www.gracesguide.co.uk/Joseph_Emerson_Dowson

Guide, G. (n.d.). *Gas Light and Coke Co*. Retrieved from Grace's Guide to British Industrial History: https://www.gracesguide.co.uk/Gas_Light_and_Coke_Co

Henderson, W. O. (1975). *The Rise of German Industrial Power 1834–1914*. University of California Press.

Henry, J. (1861). *Smithsonian Annual Report*. Smithsonian Institution.

Hirst, L. L. (1945). *Methanol Synthesis from Water Gas*. John Wiley & Sons.

Historical Coal Data. (n.d.). Retrieved 2022, from Department for Business, Energy & Industrial Strategy: https://www.gov.uk/government/statistical-data-sets/historical-coal-data-coal-production-availability-and-consumption

History of Manufactured Fuel Gases. (n.d.). Retrieved from Mikipedia: https://en.m.wikipedia.org/wiki/History_of_manufactured_fuel_gases

History of the Institute. (n.d.). Retrieved Oct 2022, from Institute of Physical Chemistry and Electrochemistry: https://www.pci.uni-hannover.de/en/institute/about/

Howard, H. C. (1945). Direct Generation of Electricity from COal and Gas. In H. H. Lowry, *Chemistry of Coal Utilization* (p. 1568). New York: John Willey & Sons.

Hunt, C. (1884). Gaseous Furl Applied to the Heating of Gas Retorts. *J. Soc Chem. Ind. Vol. III* (p. 89). Manchester: Emmot and Company.

James Prescott Joule. (n.d.). Retrieved Aug 2022, from New World Encyclopedia: https://www.newworldencyclopedia.org/entry/James_Prescott_Joule

Julius R. Mayer. (n.d.). Retrieved Aug 2022, from Oxford Reference: https://www.oxfordreference.com/display/10.1093/oi/authority.20110803100142303;jsessionid=3575C425EB97CAA54AC04E7296D0F676

List of towns and cities in England by historical population. (n.d.). Retrieved from Wikipedia: https://en.wikipedia.org/wiki/List_of_towns_and_cities_in_England_by_historical_population#cite_note-11

London, C. o. (n.d.). *Chartered Gaslight and Coke Company.* Retrieved 2021, from https://discovery.nationalarchives.gov.uk/details/r/3c478014-712e-43c6-8786-c3b871813664

Lurgi Dry-Ash Gasifier. (n.d.). Retrieved Jan 2023, from NETL/USDOE: https://netl.doe.gov/research/coal/energy-systems/gasification/gasifipedia/lurgi

Ludwig Mond. (n.d.). Retrieved from Wikisource: https://en.wikisource.org/wiki/Dictionary_of_National_Biography,_1912_supplement/Mond,_Ludwig

Manufactured Gas. (n.d.). Retrieved from USEPA.org: https://semspub.epa.gov/work/02/206912.pdf

Morris, P. J. (2015). *The Matter Factory A History of the Chemistry Lab.* Reaktion Books Ltd.

National Conservatory of Arts and Crafts, Paris. (n.d.). Retrieved from http://www.arts-et-metiers.net/musee/moteur-gaz-du-premier-type-de-lenoir

NETL. (n.d.). *Great Plains Synfuels Plant.* Retrieved Feb 2023, from US NETL: https://netl.doe.gov/research/Coal/energy-systems/gasification/gasifipedia/great-plains

NETL. (n.d.). *History.* Retrieved Feb 2023, from NETL: https://netl.doe.gov>about>history

Phillips, D. (1911). The Use of Producer Gas in Texas. *Bulletin of UT Texas, No 189,* p. p59.

Smith, R. (n.d.). *A short history of H2S.* Retrieved march 2023, from American Scientist: https://www.americanscientist.org/article/a-short-history-of-hydrogen-sulfide

The Great Peruvian Guano Bonanza: Rise, Fall, and Legacy. (n.d.). Retrieved July 6, 2022, from Council on Hemispheric Affairs: https://www.coha.org/the-great-peruvian-guano-bonanza-rise-fall-and-legacy/

Today In Science History. (n.d.). Retrieved May 13, 2021, from Gas Light Co. of Baltimore: https://todayinsci.com/Events/Technology/GasLightCoBaltimore(1881).htm

T. E. (n.d.). (1981). *1981—Wasp on Coal Slurry.*

The Annals of Tipton the growing town. (n.d.). Retrieved August 2022, from History of UK: http://www.historywebsite.co.uk/

Trewby, F. J. (n.d.). *Beckton Gas Works.* Retrieved March 2022, from Engineering timelines: http://www.engineering-timelines.com/scripts/engineeringItem.asp?id=1297

US Army Balloon Corps. (n.d.). Retrieved from Wikepedia: https://en.m.wikipedia.org/wiki/Union_Army_Balloon_Corps

UGI Corp History. (n.d.). Retrieved from Fundinguniverse.com: http://www.fundinguniverse.com/company-histories/ugi-corporation-history/

Voltaic gaseous battery. (n.d.). Retrieved July 2022, from Fuel Cell Technology: https://www.freeen ergyplanet.biz/fuel-cell-technology/the-gaseous-voltaic-battery.html

Wikimedia Commons. (2018). *File:Lenoirmotor.jpg.* Retrieved June 2023, from Wikimedia Commons: https://commons.wikimedia.org/wiki/File:Lenoirmotor.jpg

Wikipedia Commons. (2018). *File:Professor (Thaddeus S.C.) Lowe's mammoth balloon, CITY OF NEW YORK, as she will apepar when fully inflated LCCN2007681717.jpg.* Retrieved June 2023, from Wikimedia Commons: https://commons.wikimedia.org/wiki/File:Professor_(Thaddeus_S.C.)_Lowe%27s_mammoth_balloon,_CITY_OF_NEW_YORK,_as_she_will_apepar_when_f ully_inflated_LCCN2007681717.jpg

Wikimedia Commons. (2022). *File:William Murdoch's House—Redruth—geograph.org.uk— 3755601.jpg.* Retrieved June 2023, from Wikimedia Commons: https://commons.wikimedia. org/wiki/File:William_Murdoch%27s_House_-_Redruth_-_geograph.org.uk_-_3755601.jpg

Wikipedia. (n.d.). *File:Œuvres de Lavoisier Paris, Imprimerie impériale, 1862–93 RGNb10341936.22.vol III.plate IX.tif.* Retrieved June 2023, from Wikipedia, the free encyclopedia: https://en.wikipedia.org/wiki/File:%C5%92uvres_de_Lavoisier_Paris,_Imprimerie_ imp%C3%A9riale,_1862-93_RGNb10341936.22.vol_III.plate_IX.tif

(1886). *Special Report on Water Gas.* Philadelphia: Franklin Institute.